A VOYAGE ACROSS AN ANCIENT OCEAN

ALSO BY DAVID GOODRICH

A Hole in the Wind

A VOYAGE ACROSS AN ANCIENT OCEAN

A BICYCLE JOURNEY THROUGH THE NORTHERN DOMINION OF OIL

DAVID GOODRICH

PEGASUS BOOKS
NEW YORK LONDON

A VOYAGE ACROSS AN ANCIENT OCEAN

Pegasus Books Ltd.
148 W 37th Street, 13th Floor
New York, NY 10018

First Pegasus Books edition August 2020

Interior design by Maria Fernandez

Interior maps by Lara Andrea Taber

Library of Congress Cataloging-in-Publication Data is available.

ISBN: 978-1-64313-446-8

10 9 8 7 6 5 4 3 2 1

Printed in the United States of America
Distributed by Simon & Schuster
www.pegasusbooks.us

For Concetta,
who is always at the end of my roads

CONTENTS

Prologue: Cowboys and Indians   ix
The Keystone pipeline and a vision on the National Mall

1. The Takeoff Roll   1
*The cockpit at Fort McMurray Airport and motivation
for the journey*

2. The Sky Over the Oil Sands   11
*The flight from Fort McMurray Airport and background
of the oil sands*

3. Apocalypse Then: The 2016 Fort McMurray Fire   23
Canada's worst insured disaster

4. Visit to a Man Camp   35
Into the boreal forest

5. "Significant Uncertainties"   43
*Perspectives on the oil industry from offshore Louisiana
and inside Washington*

6. The Magic Juice Box   53
A moment of grace on a hard road

7. The Water Protectors   63
Battles of Indigenous people on both sides of the border

8. Stewards of the Prairie   79
Oil towns and grain towns along the rail line

9. The Bomb on the Ridgeline   91
Hardisty, the town at the head of the Keystone Pipeline

10. The Birds of Saskatchewan   107
Road trip with a prairie ornithologist

11. Where the Sea Used to Be   123
*The story of the ancient ocean, how the Alberta and Bakken oil
deposits came to be, and what they mean for the future*

12. Meet Me in Moose Jaw   139
A transcontinental rendezvous

13. The Crossing   149
At the most remote US frontier, a visit to the Teacher's Lounge

14. Williston: Boomtown, USA   161
*The fracking revolution and its changes to small-town
North Dakota*

15. The Core of the Core   173
*Approaching Theodore Roosevelt National Park through the
heart of the Bakken field*

16. A Conversation with Teddy   187
What might Roosevelt think about the dominion of oil?

17. A Certain Relentlessness   203
*How the seemingly inexorable force of the fossil fuel industry
has been turned before*

Endnotes   221

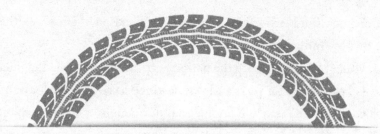

PROLOGUE
COWBOYS AND INDIANS

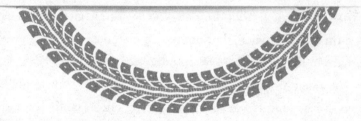

There was something dreamlike about it. For a week in April 2014, against the background of the Capitol, I saw tipis rise on the National Mall. At the end of the week, a group of ranchers in cowboy hats and Native Americans, some in full headdress, rode out on horseback onto the streets of Washington, DC, to the sound of a steady beat of drums. Beadwork, ceremonial lances, University of Nebraska vests, and plaid shirts were all in evidence. The riders led a procession of thousands to the National Museum of the American Indian to present a painted tipi as a gift to President Obama. I was tagging along for the birth of a new CIA: they called themselves the Cowboy-Indian Alliance.

What brought them to the nation's capital was an issue that had roiled the nation for years and was destined to be an issue for years to come: the Keystone XL Pipeline. For people concerned about climate, Keystone represented opening the drain to a vast reservoir of carbon in the earth, the oil sands of Alberta. The oil sands are the third largest oil reserve in the world, after Venezuela and Saudi Arabia,[1] and arguably the one that requires the most energy to produce. The purpose of this pipeline is to transport bitumen, heavy oil, from the oil sands south to refineries on the US Gulf Coast. The carbon from this reservoir would inevitably be burned to form carbon dioxide, the dominant greenhouse gas, which would remain in the atmosphere for hundreds of years.

The riders from the West cared less about the climate implications of Keystone and more about its direct effects on the land, their land. The pipeline would stretch from Canada, at the terminal town of Hardisty, Alberta, through Saskatchewan and the states of Montana, South Dakota, and Nebraska. The Canadian company TC Energy (formerly TransCanada) would acquire access to the land through the process of eminent domain. More than the disruption of pipeline construction, the westerners on the Mall were concerned about spills of bitumen. A million-barrel bitumen spill from an oil sands pipeline into Michigan's Kalamazoo River in 2010 required five years of cleanup and six years to reach a financial settlement with the pipeline company.[2] The proposed Keystone route goes across the Ogallala Aquifer, the enormous groundwater deposit beneath the US Great Plains and source of irrigation for one of the most agriculturally productive regions in the world. Air travelers from the coasts see green circles below on the golden-brown Plains, the mark of center pivot irrigation, usually from the Ogallala.

I would come to know the land these people cared about so deeply. A month after the Mall demonstration, I rode my bicycle across the Oyate Trail, the Trail of Nations, in southern South Dakota. It draws its name from the Lakota (Sioux) nations that this road connects, from the Pine Ridge to the Rosebud to the Yankton. Keystone is due to come through the Rosebud, and many of the Lakota on the Mall had come from there. I would spend my Memorial Day riding there.

It wasn't my first long bike ride. The year I retired, I rode 4,200 miles from Delaware to Oregon, looking at climate change along the way and talking with people about it. The book about that ride is called *A Hole in the Wind*. Since then, distance cycling has become a passion. But it was my background in climate science that led me to the original cross-country trek. I had spent twenty-five years working for the National Oceanic and Atmospheric Administration (NOAA), including three years as director of the UN Global Climate Observing System in Geneva, Switzerland.

April 2014 wasn't my first Keystone demonstration either. Back when I worked in climate science, our job was to lay out the facts, to describe how climate was changing and what was causing it. I didn't advocate for any policy actions, either publicly or as a private citizen. Yet the human fingerprint on climate was unmistakable, in particular the burning of fossil fuels. I had been part of the large scientific community that highlighted climate change as a problem. Now I wanted to be part of the solution. In the autumn of 2011, after returning from my retirement ride, I helped carry a giant plastic pipeline around the White House in a march along with 15,000 others in a Keystone protest. Luminaries of the climate movement like activist Bill McKibben of 350.org and NASA climate scientist James Hansen spoke to the gathered crowd.

The Cowboys and Indians procession was different than other climate demonstrations. It wasn't a bunch of stereotypical environmentalists. It harkened back to the American lore of the frontier, transcending the usual Red State–Blue State divide. The earlier political pushback had been furious. During the 2012 presidential campaign, Mitt Romney stated "I will build that pipeline if I have to myself." But there were hints that these riders from the West might be making a difference. Standing out in a suit and tie on a hot April day, I saw John Podesta, then in the Obama Administration, walking along with the march. Some days later, Jane Kleeb, one of the march organizers from the group Bold Nebraska, got an email from the White House literally saying "You've got our attention."[3]

Action was not immediate, but it came. The new Republican Congress in 2015 passed, as one of its first actions, a bill that would require approval of the Keystone XL pipeline. President Obama vetoed it. Then later that year, on November 6, he rejected the permit that the Keystone pipeline would need to cross the border from Canada. That decision would prove to be short-lived. Four days after his 2017 inauguration, President Trump signed an order restoring permits for the Keystone pipeline. Since the initial permit application from TransCanada in 2008, the effort to get this oil out of the ground has been unremitting. Where did this relentless pressure come from?

I wanted to find out. As I've discovered since my first cross-country ride, things look different from a bicycle seat. Gradually, the idea for the new bike journey began to take shape. I would start in the oil sands of northern Alberta, Canada, where the Keystone oil would come from. Later I would return to visit Hardisty, the place where the pipeline would begin. From there I would ride to another oil frontier, this one in the States. An oil field in western North Dakota has been a center of hydraulic fracturing, or fracking, a technology that has made the United States the leading oil producer in the world. This oil field is known as the Bakken. Oil production in the Bakken was booming at the very depths of the 2008–9 recession. The two deposits, the oil sands and the Bakken, were laid down by the same shallow sea that spread across the center of the continent millions of years ago. I would ride out of the deep forest of Canada's northlands and across the Great Plains that Canada and the United States share, on a voyage across that ancient ocean.

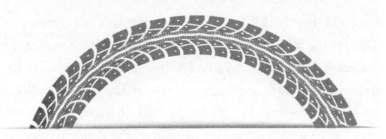

1

THE TAKEOFF ROLL

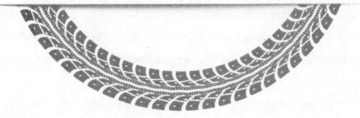

As we taxied onto the bumpy runway at Fort McMurray Airport, Matt instructed me to put on the headphones and mic so I could talk with him over roar of the engine. A pungent whiff of exhaust floated above the rows of gauges. In the distance, lightning flashed out of the bottom of a cloud.

"See that?" he said. "Another hour and I would have had to scrub on you." I wasn't quite sure whether to thank him or not.

Our little Cessna was taking off into the deep forest of northern Alberta, roughly the same latitude as central Hudson Bay. At this time of year, days last for eighteen hours and never really recede into night. Fort McMurray was the staging area and jumping-off point for the bicycle journey. I'd ride through this boreal forest and prairie south to the United States, passing over what had been the shallow sea covering much of the center of North America. Before setting off on the bike, I wanted to see this large and controversial fossil fuel project, the Alberta oil sands, source of the Keystone pipeline oil.

The scale of the operation can only be appreciated from the air. Fort McMurray is the center of the oil sands. Below Alberta's surface are 55,000 square miles of sand laced with the heavy oil known as bitumen. Open-pit mines, settling basins, processing plants, and man camps sprawl out north of town. The project is also known in the environmental community as the tar sands. Use the latter terminology anywhere near Fort McMurray and you will be corrected, in the gentle Canadian fashion.

I recalled another flight into another oil patch, forty-three years earlier. After I graduated from college, I drove down to Venice, Louisiana, on the delta of the Mississippi River, looking to work in the offshore oil fields in the Gulf of Mexico. Behind the third door I knocked on, a man looked up sullenly from behind a desk.

"How soon can you be ready to go?"

"I've got my stuff in the car."

He handed me a form. "Fill this out quick. That helicopter's leaving in twenty minutes."

It was the last I'd see of "the beach" for two weeks. I worked on the offshore oil rigs for the next four months as a roustabout, the oil field term for unskilled labor. My experience gave me some idea of why young people would be attracted to a remote outpost of the dominion of oil. It comes down to two things: money and adventure.

Different motivations brought me back to the oil patch this time around. Since my days on the Gulf, I've spent a career working in climate science. As the effects of climate change—the rising seas, the heat waves, the severe storms—grew more evident over the years, it had become clear to me that we don't have much time left to do something.

Looking at all the accumulating data was more than enough to convince me of climate change and of the role of humans in it, particularly by the burning of fossil fuels and emissions of greenhouse gases. It's not complicated. Greenhouse gases, primarily carbon dioxide, keep the sun's energy from radiating back into space, the same way that the windows of your car act to keep heat inside on a sunny day. The sources for these gases are easy to find in things that we all use or find useful: furnaces and tailpipes, jet engines and power plants. But ultimately the source of climate change—the fuel, the carbon—comes from the ground. After I retired from NOAA, I wondered: Where could I see that? Was it possible to go to where climate change comes from?

Fort McMurray is one of the biggest oil projects on the planet, and one of the most remote. It was a place to learn directly about a primary source for our changing climate. It also represented a way to satisfy my odd obsession with bike travel. Not so odd, perhaps—distance cycling brings me close to the land and to discoveries undetected from a car. Behind a windshield, you can't hear that the distress call of an antelope sounds like the caw of a crow or that the wind through a wheat field makes a low hiss.

3

Another advantage is that a solo cyclist is both unthreatening and approachable, leading to quite a few lunch-counter conversations. A bicycle can be a story machine that way. Each year since my retirement has brought another long ride and a new reason to get back in shape. Another chance to confront Danny Glover's *Lethal Weapon* line: "Maybe I'm getting a little old for this shit."

Fort McMurray would also allow me to return to thinking and writing about climate again. I was fully wrapped up in the idea one day when my wife, Concetta, asked me what was on my to-do list.

"Save the world, write the great American novel, and fix the upstairs toilet," I said.

A sidelong look ensued. "I'd settle for one out of three," she said.

Still, the idea of riding to where climate change comes from stuck in the back of my head. Never completely sold on the idea of a solo ride through the wilderness, I set about recruiting a riding companion. It proved to be a little more difficult than I imagined. I'd had a number of stalwart companions on earlier rides, but for some reason they all seemed to be coming up with really important places they needed to be in a year's time. Jan Kublick and I had found Hemingway's piano as we rode on El Camino de Santiago across Spain. A willingness to ride long distances for esoteric ends made him a logical candidate. But when I brought up the ride to him, the response was not what I'd hoped for.

"Alberta? But not the Rockies?" He tried to bring me down gently. "Not much in the way of wine up there."

"Yeah, but I'm sure they've got some great craft beers."

"Seriously? You'd go for beer brewed from the tar sands?"

I was getting the idea this might be a tough sell. Rick Sullivan and I had explored juke joints in the Mississippi Delta on a ride north from New Orleans. He was a little more direct.

"So why exactly would I want to go to the shit show?" he asked. This was looking more and more like a solo ride.

A few other nagging doubts about the trip began to accumulate. I was heading into the deep dark forest of northern Canada, and I could find no guidebooks or blogs about riding in this part of the world. However, a friend helpfully sent a link to an article about a grizzly coming over the guardrail to chase a cyclist in British Columbia. Meals on wheels indeed. I began to envision the boreal forest as populated with toothy creatures intent on tracking me down. Realistically, the threat was more likely to be big oil trucks and narrow shoulders. Considering my background in climate science, some of the people driving those trucks might not exactly take a shine to me. It's a really quiet road. A little bump and no one would find me for a long time.

But I was still intent on seeing where carbon comes out of the ground. In terms of carbon dioxide released per barrel of crude, Alberta's oil sands produce some of the dirtiest oil in the world.[1] It's not hard to see why. For oil sands mining, producing one barrel of crude oil means that about two tons of sand must be dug up, moved, and processed with about three barrels of water.[2] Heating the water, adding chemicals, and moving it all around uses a tremendous amount of energy, up to a third of the energy content of the bitumen.[3] Consider how much energy it takes to melt tar in the middle of an Alberta winter. Heating water means burning things, namely oil and natural gas, adding still more carbon dioxide to the atmosphere.

The oil sands and the process of getting the oil to market have churned politics, and not just of the United States. Climate scientist James Hansen referred to the Keystone Pipeline as "the fuse to the biggest carbon bomb on the planet."[4] People heard this in Canada as well. A proposed pipeline expansion to bring oil sands bitumen west from Alberta to Pacific ports was vehemently opposed by the provincial government of British Columbia. When Kinder Morgan, the company building the pipeline, threatened to pull out because of this opposition, the Canadian national government took the controversial step of purchasing the pipeline to try to ensure its construction. In different directions and in different nations, the oil sands bitumen is seeking to find its way out of the ground, powered by immense financial interests. I wanted to see this giant project in the forest up close, directly. To bear witness, and hopefully not witness bears.

Oil industry people in Fort McMurray would argue, correctly, that even more new carbon is coming out of the ground in the United States, because of the introduction of hydraulic fracturing, or fracking. On my bike ride, I would see that too. The oil sands are far from the only massive oil field in this part of the world. On the other side of the border, in the Bakken, use of fracking has led to an enormous increase in US oil production and altered the global energy equation. In part due to Bakken drilling, the United States surpassed Russia and Saudi Arabia in 2018 as the largest global crude oil producer.[5] At the bottom of the recession in 2008, the lowest unemployment rate in the United States was in North

Dakota. Williston, the largest settlement in the Bakken, became the States' newest boomtown. In 2013, analysts of nighttime satellite images began to notice a mysterious patch of light on the high plains of North Dakota, bigger than metropolitan Chicago.[6] The mystery didn't take long to solve. The light was coming from natural gas flaring off thousands of wells.

In the Bakken, what comes up out of the wells besides oil and gas is salt water, leftover brines from that shallow sea that covered much of inland North America millions of years ago. The Alberta oil sands and the Bakken are both remnants of this ancient ocean. They are two holes in the earth from which carbon deposited during those epochs now finds its way back into the atmosphere.

These two massive fossil fuel projects highlight some of the planetary-scale effects of oil and gas operations. A tagline in a *Business Insider* photo essay on the oil sands crystalized the issue:

> Despite how you feel about the environmental impact oil companies may have on the world, they're not likely to go anywhere while there is still oil left to collect—with or without the Keystone XL pipeline.[7]

That pretty well summarizes it. Fossil fuel companies, left to their own devices, will extract every profitable drop of oil they can find. Christophe de Margerie, late chairman of the French oil company Total, once gave a terse and candid response to the question of what to keep in the ground: "I am not in charge of the planet." It shouldn't surprise anyone. For all the cheery ads about saving butterflies and growing algae, oil companies are mostly about getting fossil fuels from the ground. But extracting and burning all of these huge stocks of oil and gas will lead, by any reasonable interpretation of the science, to an increasingly warm and less habitable planet. Our best estimates are that roughly 80 percent of currently known stocks of fossil fuels need to stay in the ground to keep climate change within somewhat manageable limits.[8]

We have what appears to be an unstoppable force, the fossil fuel industry, hurtling us toward a mortal threat to the habitability of large parts of the planet. What is the nature of this great force? As I've found everywhere I've looked, things are different at ground level. I wanted to have a look, to ride between Fort McMurray, Alberta, to Williston, North Dakota. Surely there are nicer places to ride. Some of the appeal was, to paraphrase John F. Kennedy, that I would ride not because it's easy but because it's hard. I had the sense that to see these places and describe them was the most important thing I could do.

I set about planning a ride across this ancient sea, to explore its people and places and animals. The journey, from deep forest to prairie to badlands, Canada to the United States, Alberta to Saskatchewan to North Dakota, would stretch 1,100 miles. It's even more impressive in kilometers. For the entire ride, the dominion of oil would never be far away.

The route is the stuff of deep winter dreams, or nightmares perhaps. In the cold-weather months, planning involved windows of maps and satellite views and motel web sites splayed across and piled onto my computer screen, mirroring the mess of my office. I sought a bicycle route through the boreal forest ending each night with a roof over my head so I wouldn't have to carry camping gear.

The tradeoff for that lighter load was that each day would be committing: I would need to reach that day's destination and bed, turn back, or spend the night in a ditch. The route that formed on those late nights would entail at least one night in the man camps, the massive prefab lodges built to house oil field workers. It would also involve seventy-five-mile days with few or no settlements along the way. That would be close to the limit for sixty-five-year-old legs, or mine anyway.

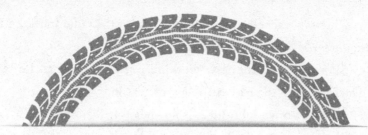

2

THE SKY OVER THE OIL SANDS

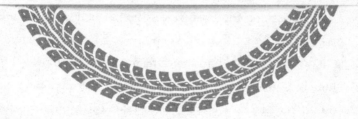

Matt was busy as we climbed out of Fort McMurray Airport, sturdy hands on the yoke, talking with the tower, announcing our presence to other aircraft. The air was bouncy and unstable. Our ascent revealed billowing storm clouds, giants towering in the distance, their dark undersides spitting occasional lightning. Matt had a three-day beard and a quiet, confident air. He had seemed almost too big to fit in the cockpit of the four-seater Cessna. When I asked to take his picture before we took off, he instinctively leaned on the front of the plane, fond of this piece of metal that took him into the sky and back every day.

When we reached altitude, I asked about the radio traffic crackling in what I'd thought would be a quiet place.

"They've got a fire crew working out to the west," he said. "Doesn't sound too bad. Still, it's the season."

"But most of your business is for the oil fields, right?"

"There's some, yeah. But a lot of these little towns up north don't even have roads in. This time of year there are lots of hunting and fishing parties."

"How'd you get into being a bush pilot?"

"I pretty much grew up in the cockpit. My uncle flew—still does—and he would take me up all the time. Just came natural."

I looked back on the Fort Mac airport. I had landed there the night before, after a trip from my home in Maryland. Before that long flight to Canada, I had spent the previous day in my garage in the June heat, disassembling and packing up my Trek bike, trying to pad all the places where it might take a lick on airport carts and conveyors. Like Matt and his Cessna, the bike and I have a certain bond, having come through a dozen trips of 500 miles or more. There's a little rust spot on the top tube two feet under my chin where my sweat drips. Maybe someday that's where the frame will break, but for now it's a mark of how the bike and I feel like a single entity sometimes. The old Trek had been tucked snugly into the belly of the Air Canada flight.

The night before the flight with Matt, my jet arrived into Fort Mac well after midnight. We flew over much of the route I'd be riding. I pondered the ancient ocean a few miles below in the persistent twilight. Summer solstice was just a few days past, and with Fort McMurray at latitude 59° north, darkness was never complete. The Trek and I reunited at baggage claim. I hauled my gear out to a lonely taxi stand, where I had been assured that a van would be forthcoming. Mosquitoes swarmed out of the half-light. I needed more than two hands to keep them off. Welcome to the boreal forest.

I was back in the air the next day in the more intimate confines of Matt's cockpit. When we leveled off, he became a part-time tour guide. He pointed me to the city of Fort McMurray below. Before returning to the airport, I'd had the chance to wander through the downtown. It's a clean, modern city of 80,000, with all of the amenities you'd expect: shopping center, library, community college, golf course. In short, not much that would reveal how remote it is. The nearest city of any size is Edmonton, some 270 miles to the south. Yellowknife, in the Northwest Territories, lies 410 miles to the north. The Athabasca River flows north through town on its way to the Arctic Ocean. The view from the cockpit revealed one startling aspect of Fort McMurray: On all sides, it's surrounded by fire-blackened trees.

Matt motioned for me to look ahead. In the distance, the earth opened up. What I was seeing can only be described as moonscape. The trees had been removed and piled next to a lumber mill on the nearby Athabasca River. The face of the land was one of terraced, open-pit mines, conveyor belts hauling oil sands, and massive processing plants called upgraders. Water was everywhere, from puddles in the mines to rivers to lakes, both naturally formed and constructed for waste disposal. On the mine face, massive orange shovels took multi-ton bites of oil sand and deposited them into fleets of waiting trucks that dumped the sand into giant vats.

The stream of trucks moving from the mine face appeared like lines of ants moving on an anthill, but the trucks are anything but tiny. In fact, they're the biggest in the world. The tires alone cost $60,000 apiece. Matt pointed down to one string of trucks.

"I know a guy who runs one. He says it's like driving your house from the upstairs bathroom window."

A white structure with dozens and dozens of wings sprawled near the river. The lodge, better known as a man camp, was built to house the hundreds of workers that run the mines. Each wing of the camp has twenty-three rooms. I'd have a chance to see one up close in a few days.

Most evident from the air were the tailings ponds. Before coming to Fort McMurray, I'd seen them in the satellite view on Google Maps. "Pond" is a serious euphemism for some of the largest man-made structures on the planet. Separating oil from sand creates, as a byproduct, a mixture of water, fine silt, and largely toxic industrial solvents called tailings that initially float on the surface of these giant lakes. It takes years for the materials to settle out. The scale is staggering. Vast pools of oil and chemicals swirl across the surface off lethal shorelines.

This bitumen froth has a strange beauty to it. Matt and I flew over a tailings pond that from the air resembled the mottled brown-and-white foam in a latte. A 1942 report on an early separation plant stated, "Tailings disposal will require a large area." By 1989, the tailings ponds were larger than the city of Fort McMurray.[1]

The issues with tailings disposal had been highlighted on a snowy day in April 2008. It was the season for waterfowl migration.

Flocks of mallards and mergansers were on the move north to their summer feeding grounds in the Peace–Athabasca Delta just as a late season blizzard was subsiding. Tailings ponds are normally surrounded with propane-fueled sound cannons to keep birds away. The sound cannons weren't working on the Syncrude Aurora Settling Basin that day. The open water, or what looked like water, must have been a tempting place to land. A total of 1,611 ducks went in. Five came out. Photographs of lumps on the surface of the tailings pond were on the front pages of newspapers around the world.[2]

Matt banked right at the northernmost point of our flight. He pointed out a small community below us.

"That's Fort McKay. Wouldn't live there," he volunteered. "Just some nasty smells."

He wasn't exaggerating. Oil sands mines and upgrading plants are far closer to the First Nations community of Fort McKay than to the larger Fort McMurray further south. That summer in Fort McKay saw the introduction of an app to report odors coming off the oil sands. As one resident said, "The sulphur smells like rotten eggs, and the ammonia smells like strong cat pee."[3]

As we turned south, back to Fort McMurray, we passed over more open-pit mines, and one tailings pond after another. Boats glided across the larger ponds, dragging orange booms out from the industrial shoreline into the oil sheen, separating the bitumen from more open water parts of the giant lake.

Green began to reappear below us, but not quite how I was expecting. For miles and miles, what looked to be regrowth of grass and bushes lay below blackened tree trunks.

Amid a landscape of refineries and roads, open-pit mines and heavy equipment, what caught my eye were bright, square, golden slabs. One of the byproducts of the processing of oil sands is sulfur, far more than anyone can use. Below us, on the property of the Syncrude plant north of Fort McMurray, monumental gold structures, poured from elemental liquid sulfur, dominated the landscape. One calculation puts them, by both base area and volume, as larger than the Great Pyramid of Giza.[4] Everything about the oil sands is huge.

The thunderheads moved closer as Matt turned to line us up on the runway. He set the wheels down gently on the tarmac and taxied over to McMurray Aviation. He chocked the wheels for the approaching storm as I was unlocking my bike. We shook hands, and I managed to get in the door of my next stop before the squall line swept across.

The Oil Sands Discovery Centre is a big, modern museum set up by the province of Alberta and dedicated to telling the story of the oil sands as a Canadian national project. The narrative is of "doers and dreamers" looking to harness this vast oil deposit. Working the oil sands is anything but new. One of the first patents on the hot-water process used to separate oil from the sands was obtained back in the 1920s. After many false starts, production really began to take off in the early 2000s.

By 2006, Alberta's employment rolls were adding new jobs at a pace of ten thousand a month. Fort McMurray became a full-scale boomtown. The newly employed were living in campers in hotel parking lots, renting out backyard toolsheds, and sleeping in tent villages on the outskirts of town through a subarctic winter.

Meanwhile that same year, the Smithsonian Folklife Festival in Washington, DC, featured the culture of Alberta. A Caterpillar haul truck from the oil sands with its two-hundred-ton dump bed was parked on the National Mall.[5] Alberta was in full swagger mode.

The oil business is notorious for its booms and busts, and the oil sands are subject to the same swings. An oil price crash in 2014 brought Fort McMurray back down to earth. Yet another in 2020, driven by the coronavirus outbreak and a price war in the global oil markets, has oil prices below the break-even point for most Canadian producers.[6] Still, Canada remains a major international oil exporter. For all the discussion of the Middle East, the United States gets more foreign oil from Canada than anywhere else.

The last section of the Oil Sands Discovery Centre is labeled "Environment." It features a pipeline with a cheery cartoon voice emerging from its depths to explain how oil gets to market. A tagline on the last panel states, "When we use energy from the oil sands to make our lives more pleasant, it has an impact and an environmental cost." Just what that cost is cannot be found in the Discovery Centre. But outside, framing the entrance to the Centre, are two matched pairs of charred trees.

Late the next afternoon, I waited at a window table of the Fort McMurray McDonalds. Though we'd never met, I recognized Bryan immediately from the photo he'd sent. His expression had a certain openness, and he had big, broad shoulders—just what would you expect from a power engineer. Our meeting was both random and fortuitous. He had encountered two cycling friends of mine on a trip to British Columbia. They mentioned that I was a writer visiting Fort McMurray by bike. He told them that he could take me around the oil sands operation. In later texts to me, he said that he "could offer a different perspective about the oil sands that might not be as familiar as what you see in the media." That was how he came to be stepping down from his SUV in the McDonalds lot.

I introduced myself as he came in the door. He had a strong handshake and a direct way of talking. Bryan isn't his real name, as he had some unauthorized stories about the oil patch.

"I understand you're from here," I said.

"Born and raised," he nodded. "Grew up with the town."

"I appreciate your carving out the time, and I know it was a stretch for you to make it here." We'd had a couple of reschedules. "I've met a few folks working in town, but you're the first from the oil patch proper. Could you tell me just what a power engineer does?"

"Well, I've worked a lot with the boilers that liquify the bitumen up here. Water and steam and chemicals moving around keep this place going. It takes a lot to get the oil out up here. Not just sticking a straw in the ground and sucking it out like they do in Saudi Arabia. But we've still got some good daylight. Want to take a ride up to Syncrude?"

"Love to."

The shadows were lengthening as we rolled north through Fort McMurray to the open pit mining areas, part of what I'd flown over the day before. The roads were filled with pickups, many with poles on the back known as "buggy whips" to alert drivers of the much larger oil sands trucks to their presence. This is indeed the land of trucks. The local brewery even has a Lift Kit Lager.

Our first stop was next to a major tailings pond. Up a hill overlooking the pond was the Bison Viewpoint, a viewing area for wood buffalo that's part of land reclaimed from the mines. The buffalo weren't far, well fenced off from the tailings pond, but the mosquitoes chased us away in the low light. Not far up the road was Gateway Hill, a forest that was also once an old mine. Still further on, Bryan pointed to what looked like a giant sand pit.

"This is an old mine site where they're working on reclamation. They use sand left over from the already processed oil sands. They add nutrients and replant, eventually returning the mine to as natural a type of place as can be, considering what the area went through to get there."

I had my skepticism. From the air, Gateway Hill and the Bison Viewpoint appear as tiny green patches in a sea of open-pit mines and tailings ponds. The oil companies' most optimistic accounting is that about 11 percent of the area disturbed has been or is being reclaimed.[7] But local ecologists assert that the intensive disturbances change fundamental biological processes, making it impossible to fully restore the affected lands. They maintain that the principal aim of these reclamation projects is to get the social license needed for the companies to continue operating.[8] The companies have certainly expended a lot of effort to produce these forests and grasslands. While the area of the reclamation projects was quite small, they were directly adjacent to the road.

In 1787, Russian Empress Catherine the Great went on a tour of newly acquired regions of her empire. The story goes that for her benefit, minister Grigory Potemkin had small portable settlements built along the banks of the river before she went by. Ever since, a construction built to convince onlookers that the situation is better than it is goes by the name of a Potemkin Village. Driving past Gateway Hill and the Bison Viewpoint, my sense was that we might be looking on a Potemkin forest.

Our next stop was an industrial plant, a maze of pipes and towers, what looked like a refinery to my untrained eye. Bryan explained it for me.

"This is the Syncrude upgrader facility. Spent a good bit of time here. When Exxon bought Syncrude, they did a lot of cost cutting. Brought in all the real serious oilmen from Houston." He gave a little wry smile. "But they just didn't know how to handle forty-below winters. Ice castles are all over the plant. We had these little heaters all around that were keeping the pipes thawed. They cut them out as a cost saving measure. Quite a mess. Safety just went down the tubes."

Bryan looked up at the stacks of the plant. "I thought I was going to show you a lot more going on. But it doesn't look like the plant's running right now."

He was right. A transformer explosion had brought the Syncrude plant down for several months. Some measure of the importance of this operation is that the ripple effects had increased global oil prices. According to *Bloomberg News*, the Syncrude outage influenced prices more than a recent OPEC action.[9]

Not far from the upgrader, Bryan pulled up to an active tailings pond, the rainbow sheen on the surface glistening behind a chain-link fence. In the gathering twilight, I heard what at first sounded like the thump-thump of distant artillery, as though we were on some kind of battlefield. The pounding was the report of propane cannons sounding continuously across the toxic landscape. Between the blasts, speakers broadcast recordings of raptor calls. Hawk silhouettes were posted on pilings out in the water. The companies were clearly taking great pains to discourage birds from landing. The site of the duck disaster was made to appear just as deadly for waterfowl as it was.

On the way back into town, I noticed a notch on the horizon. I thought it was the road out of town.

"Looks like that's where I'm headed," I said.

Bryan tapped on his glasses. "Might be time for a stronger prescription, my friend. That notch is a transmission tower, connecting the Alberta grid to the rest of North America. A lot of that electricity ends up in California, so LA is one of the biggest consumers of oil sands electricity. Something I'm sure Darryl Hannah might turn a blind eye to when it comes to lighting up her mansion at night."

Bryan's was a familiar and valid refrain in the oil patch: We're doing a dirty, dangerous job producing stuff that everyone uses. There's no trace of apology in his manner. In a later email, he would write:

> Stay in touch and please spread the word that although we're from the oil sands, we're not bad people. We all want a better future for the planet and to leave it greener than we found it, and we're trying our best to learn

21

better and more efficient ways of processing the oil sands every day. We're all just kind of caught between a rock and a hard place with having to pay bills, put food on the table, and eventually retire.

Bryan and I surely disagree on many things. On one side of that disagreement lies the future of the planet, the loss of many coastal cities, vanishing habitability for significant areas, and a reshuffling of the evolutionary deck. On the other side lies the future of Bryan's livelihood, and that of Fort McMurray, the province of Alberta, and a fair piece of the world. In many ways, this exemplifies the climate change problem. The benefits of extracting fossil fuels are here and now, and they keep Bryan's home town afloat. The environmental costs are global and will be with us for thousands of years.

We drove back toward Fort McMurray on a ridge above town. Bryan had a certain wistfulness as he pointed to the town below.

"We used to have some monumental traffic jams in town back ten years ago. It's not such a problem now." I'd seen some boarded-up storefronts along Franklin Avenue, Fort McMurray's main drag. "And the aspen around town would turn yellow in the fall. They're all gone now."

We turned into Timberlea, Bryan's old neighborhood. The landscape was of delivery trucks, piles of construction material, half-framed structures: homes, town houses, apartments in various stages of completion. He pointed out the fire scars on the siding of one house. He began to spin the tale of the great fire of 2016. Around this part of the world, they call it The Beast.

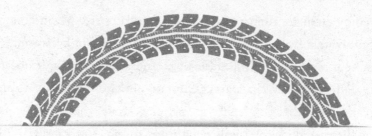

3

APOCALYPSE THEN: THE 2016 FORT MCMURRAY FIRE

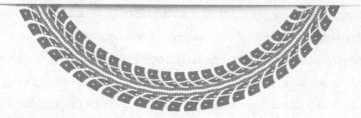

When I was growing up, Smokey Bear was all over TV. He was a big, deep-voiced bear, urging you to douse campfires and snuff out matches. He would point his finger at you from the screen, closing the commercial with his catchphrase: "Only YOU can prevent forest fires." I believed absolutely. If you can't trust a bear in a ranger hat, who can you trust?

Smokey lied. Well, not completely. But the origin of many fires has nothing to do with humans. In Canada, lightning starts 45 percent of all fires, and those fires represent 81 percent of area burned.[1] Still, when it gets hot enough, and dry enough, and windy enough, almost anything will start a fire. Maybe one of Smokey's misplaced matches, or heat from a motorcycle muffler, or sparks from a railroad. Or lightning. The point is, it doesn't really matter how it starts. Climate change is getting things plenty warm. If you heat popcorn on a stove, there's no way to predict which kernel will get things going. But one will. On May 1, 2016, in the heat and the drought, something popped 12 miles west of Fort McMurray.

What preceded that little wisp of white smoke is anyone's guess. The conflagration that ensued became variously known as the Horse Creek Fire or the Wood Buffalo Fire or Fire 9. In Fort McMurray, they just know it as The Beast. But back on May 1, the white smoke indicated a grass fire burning along the right-of-way of a power transmission line. Two other fires in the area had been snuffed out in the previous two days. The fire crew on the Bell 212 helicopter figured they'd snuff this one out too. They landed in a clearing minutes after they had called in the fire to the control center.

This fire was different. The crew on the ground noticed that the smoke had changed from white to black. The grass fire had leapt into the bone-dry trees and into the treetops before they could cut it off. The firefighters didn't have enough hose to reach the stream. The crowns of the trees, conifers and aspens, began to explode as the fire advanced at thirty feet per minute. It was the firefighters' worst nightmare: a running crown fire. The helicopter crew retreated. The Beast was out of control.

The battle was now up to the big airborne tankers. But their first fire-retardant drop went to extinguish another fire inside the Fort McMurray city limits. Maybe they went after the wrong fire. More than likely, nothing would have mattered. It was just too hot and the forest was too dry.[2]

The next day the fire grew, but the wind didn't blow it toward the city. It left a cigar-shaped burn area aimed at the suburbs to the south. Fire crews desperately tried to cut a line of defense through the forest between the Beast and Fort McMurray. But it remained a big, hot, dangerous fire. The fire chief warned, "The fire is able to get into areas where we can't stop it."

The lull ended around noon on May 3. The wind picked up, fanning hot spots and once again driving flames into the treetops. Embers blown high in the air started new blazes over the fire lines. The fire jumped the Athabasca River, and it now had a clear path to the northern suburbs. In all of the firefighter's training exercises and emergency scenarios, the great expanse of the river was supposed to be the hard barrier. Not this day. Towers of fire turned toward the city and came sweeping out of the west. The Beast was now moving on Fort McMurray.

An hour before that, Alberta Agriculture and Forestry and the Wood Buffalo Region held a joint press conference, warning ominously that fire conditions were extreme and urging residents to be "prepared to act on short notice." However, to instill calm, the authorities also told residents to "get on with their lives and take their kids to school." Many residents did.[3] For the children at the Good Shepherd Catholic School in the Beacon Hill neighborhood, that was a bad idea. By 1:30 P.M., a wall of smoke rolled across Fort McMurray, blotting out the sun.

Bryan, who had driven me through the oil sands, described what happened that afternoon:

> At the elementary school, kids looked out the window and the fire was right there, right at the forest edge. The teachers didn't have time to call parents. They just started throwing kids in whatever cars or buses were there to get them out. The parents were going crazy because they couldn't find their children. But every single one of them made it out.

In another part of the same Beacon Hill neighborhood, a phone call from a friend awakened Michel Chamberland, an oil field worker, in mid-afternoon after his night shift. He opened the door to see the fire on the other side of the street. Chamberland captured the evacuation of the Beacon Hill neighborhood from the dashcam on his pickup. The images are of Dante's Inferno with brake lights. In the midday darkness, only exploding trees provided light. Cinders splashed on the hood. Watching the video, you can almost feel the heat through the window of his truck. Flames washed over the top of his roof, scorching the paint. The traffic was remarkably orderly, in that Canadian way. Drivers stopped for red lights. They allowed others to merge. The soundtrack is the background roar of fire outside the windows. Chamberland can be heard whispering: "My God . . . shit . . ." Embers swirled on the road. A police officer in shirtsleeves directed traffic.

This was part of the full-scale evacuation of Fort McMurray. Firefighters went door to door, urging residents to get out. Many insisted that they needed to gather possessions or wait to meet someone. As embers rained down, firefighters were firm: "You're out of time." Gas to the city was cut to avoid explosions, but that also knocked out the gas-fueled pump for the fire hydrants. They ran out of water.[4]

On the way out of town, air tankers dropped loads of fire retardant to provide cover to the lines of vehicles jammed in traffic. Incredibly, only two of eighty-eight thousand died, and those in a traffic accident. This is a young place: two-thirds of the population are between the ages of twenty and fifty-five. These young residents of an oil town know safety, they know safety drills, and they know about staying calm in a crisis. This was their moment. To the south, on the road to Edmonton, families opened their homes to evacuees pouring out of Fort McMurray. In the north, the man camps around the mines became family camps.

Though thousands of houses had burned, most of the empty city remained intact by the evening of May 3. Fire crews and aircraft poured in from across the country, with equipment setting up at the MacDonald Island Park. They would make a last stand to defend the city.

The Beast was rising up, quite literally. From south of the city, on May 4, an employee of Alberta Agriculture and Forestry snapped a photograph of towering white clouds above the darkness of smoke. Powerful updrafts from the fire were punching into the stratosphere, creating what meteorologists call pyrocumulonimbus, or fire-driven thunderheads. The aircraft and the fire crews were now trying to keep a firestorm at bay.

Everyone has a story of those days. Bryan stopped to talk with me along the road in the Timberlea development, on a bluff looking out over downtown.

"It was a pretty desperate fight to save the hospital from the Beast," he said. "They came with plane-load after plane-load of fire retardant. Along with the hospital, they were trying to defend the city center. And they pretty much did, although neighborhoods were destroyed."

Matt, the pilot from Fort McMurray Aviation who had flown me over the oil sands, had his own account.

"I'm not certified to fly the big stuff, but my uncle is. He was up that day. It was quite a show," he said. "Ninety-five aircraft at the peak, tankers and helicopters, trying to keep the fire out of the city. Shut down the airport just so they could handle them all."

"Did you guys have any problems with the fire on the ground?" I asked.

He smiled a little and motioned with his head. "You see across the road?" Charred trunks still stood. "It was right on top of us here at the airport, but we got out okay. Still, a couple of pilots lost their houses."

More than just a couple of pilots were burned out. While the city center escaped, today it's an island surrounded by seared forest, with scrub-willow undergrowth greening up after two years. Much of the Timberlea neighborhood, where Bryan had driven me through, survived. Not so lucky was Beacon Hill, across the road from my motel and the start of Michel Chamberland's fiery ride. Most of it was burned out. The Good Shepherd School was damaged by smoke and closed for a year. The fire would continue burning for five weeks, destroying or damaging almost 2,000 structures and burning 1,455,000 acres, a larger area than the state of Delaware. The total economic impact was $6.7 billion, making it the costliest insured disaster in Canadian history.[5] People helping people in their time of need is the story of both the fire and the recovery. Two years on, I could see the evidence of that recovery everywhere. The town was rebuilding.

As the fire was sweeping past Fort McMurray, an information technology manager from southern Alberta posted a tweet, which was quickly deleted, that read, "Karmic #climatechange fire burns CDN oilsands city." In the storm of criticism that followed, he managed to keep his job, but barely.[6] One must tread softly when people are fleeing their burning homes.

A similar dynamic exists in the United States. Following the devastating California fires of November 2018, Interior Secretary Ryan Zinke, a principal proponent of fossil fuel extraction and "US energy dominance," toured the area. In his comments at a news conference, Zinke acknowledged the role of rising temperatures in the wildfires while emphasizing the role of forest management. The fire seasons have "gotten longer, the temperatures have gotten hotter," he said, though he also said "there's a lot of reasons for a fire" and "now is not the time to point fingers."[7]

In a disaster, it's difficult to talk about long-term solutions in the immediate aftermath, or probe the root cause. In the wake of any tragedy, climate-related or otherwise, the first task is for people to get their lives back together. For Fort McMurray, it's been a couple of years. New growth is sprouting up around the blackened trunks. Perhaps it's time to tread softly, but to tread nevertheless. So the question is: How did the hotter climate of Fort McMurray in May 2016 develop? How did we get here?

There's no question that in May 2016, the forest of northern Alberta was primed to burn. A post-incident report noted that the wildfire season had started earlier and lasted longer than it had historically. In 2016, the wildfire hazard risk in northern Alberta was severe due to extreme dryness the previous summer, low moisture over the winter, and dry, warm conditions in the early spring. These conditions reached unprecedented levels by May and were the largest contributing factor to the great fire.[8] Paradoxically, firefighting tactics also played a role. Like the western United States, Alberta suppressed natural fires that would have burned off underbrush for decades. So in 2016, there was a buildup of tinder, fuel for the blaze.

On the larger scale, climate change is the background music for fire both in western Canada and the western United States. Warmer and drier conditions have contributed to a dramatic increase in large forest fires in the western continental United States and interior of Alaska since the early 1980s, a trend that is expected to continue as the climate warms and the fire season lengthens.[9] Similar trends are present in the forests of western Canada. Natural Resources Canada states that "in Canada's northwestern boreal regions, the annual amount of forest area burned by wildland fires rose steadily over the second half of the 20th century."[10] But there's a fundamental difference between the boreal forest and the western United States. In the Arctic, warming is two to three times greater than the global average.[11]

The primary reason for this warming is the burning of fossil fuels. Carbon, coming from the ground as hydrocarbons (coal, oil, and gas) is transformed into carbon dioxide, the principal green-house gas, during the combustion process. That carbon dioxide stays in the atmosphere anywhere from decades to centuries. Absent some way of removing it from the atmosphere, the results of burning fossil fuels will be with us for generations.[12]

Can we burn the fossil fuels we know about, the known reserves, and still have some semblance of a livable climate? Not even close. International negotiations have set 1.5°C as an upper limit for how high the global average temperature should be allowed to go. The known fossil fuels reserves represent 942 gigatons of carbon dioxide. The carbon budget to stay within 1.5°C is a mere 353 gigatons.[13] The conclusion is unmistakable: To prevent the most damaging consequences of climate change, we must make a conscious decision to leave the majority of the world's fossil fuel resources underground.

Given that kind of a severely constrained carbon budget, what kind of fossil fuels should stay in the ground? A prominent 2015 study in the journal *Nature* used the not unreasonable premise of keeping the dirtiest, i.e., most carbon-intensive, fuels in the ground.[14] The study models what would be needed, given their assumption, to keep the planet at 2.0°C. Use of coal, the most carbon-intensive, would need to slow rapidly, with 82 percent of current reserves to be left unused. Not far behind are Canada's oil reserves, consisting mainly of the oil sands, which are among the most carbon-intensive oil in the world. The study estimates that about three-quarters of these oil reserves would need to be left undeveloped.

The study made headlines around Canada. The response in much of Alberta was: Why us? One writer encapsulates it:

"Why was bitumen being asked to be the first to atone for all the world's climate sins?"[15] Basically, the idea is that we (in this case, the oil sands) are just a tiny part of a global problem. It's a point made by everyone producing fossil fuels around the world. Everyone points to someone else, and passing on the responsibility is a good part of the reason that no action happens.

Back to that treading softly part: Did Fort McMurray's business have something to do with Fort McMurray's tragedy? At the Oil Sands Discovery Centre, they make the point that this is the third biggest oil deposit on the planet. It's hard to make that case while simultaneously maintaining that oil sands are an infinitesimal part of the climate change problem. Simply put, carbon, of which the oil sands are a huge source, is a primary driver of climate change, the very thing that set up the boreal forest to burn. It's not karma. It's physics.

The only way to slow the warming, the droughts, and the fires is to reduce our production and consumption of fossil fuels and to shift to renewable sources of energy. Ultimately humans hold the key to slowing this down. Only you. I think about a bear pointing out from a TV commercial. Maybe Smokey was right after all.

On my last afternoon in Fort McMurray, I rode the bike up to MacDonald Island, where the town has a beautiful park and community center on the Athabasca River. The island was one of the places that hadn't burned, where the fire crews had gathered their equipment for the epic last stand in 2016. It was a bit quieter two years on, with no firestorm or squadrons of tankers overhead.

I was at the northernmost point of my ride, or of my life, for that matter, where the river winds toward the Arctic Ocean. There was something surreal to it. I was under an umbrella having a Riesling on the patio of the golf club on MacDonald, a brief pause before embarking on the big ride the next day. On a sparkling, warm day, flowers in flower boxes were blooming at the height of the northern summer. In the foreground, golfers were measuring out their putts on the nearby green. They were framed by charred hillsides in all directions, the burnt smells washed out after several years. Out of sight beyond the hills were the man camps I had seen from the plane days before. The next evening, I would be staying at one.

As Alberta and the entire oil sands industry would attest, Canada is far from alone in contributing to the climate problem. The United States contributes almost ten times the carbon dioxide of Canada.[16] In particular, that new shale oil industry south of the border is also pulling vast amounts of carbon out of the ground. It's another piece of the global climate problem, this one on the US side of the border, about a thousand miles away. Time to ride south to go see it. Time to take to the highway.

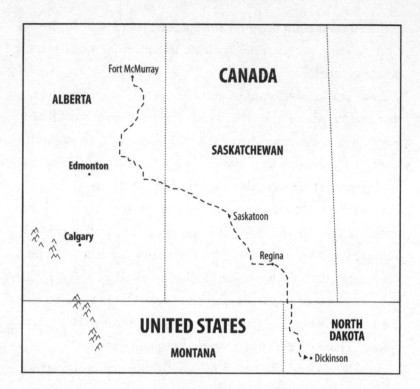

*The route: Fort McMurray, Alberta,
to Dickinson, North Dakota, 1,095 miles.*

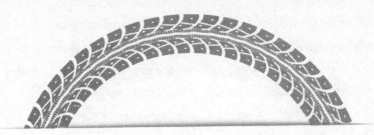

4

VISIT TO A MAN CAMP

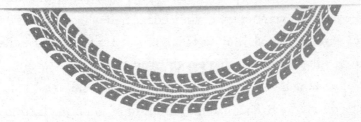

The sign on the outskirts of Fort McMurray read No Gas or Services Next 200 km. I felt a wave of nausea and resolved to ignore it. It should have been no surprise, but in that moment it was breathtakingly clear how far out there I was. The two rear bike bags were packed up tight, and I carried four bottles of water and energy drinks. I'd tried to be prepared for common and uncommon breakdowns on the bike. This was what all the months of planning on gray winter days had been for, but my padded office chair and colored satellite maps seemed very distant. Ahead were five days of brutal riding into the boreal forest.

South from Fort McMurray lay the most remote leg of my ride, into the dark woods and a different kind of oil country. The nearest motel was 95 miles down the road, too long a stretch for the first day's ride. If I wasn't going to end up spending the night in a roadside ditch, I would need to rely on a "lodge" adjacent to a drilling facility—a man camp.

The boreal forest that I would ride through circles the Northern Hemisphere with a ring of deep green, primarily coniferous trees. It stretches across Alaska, Siberia, and Scandinavia. Canada's piece of this emerald ring includes a quarter of the world's intact (i.e., non-fragmented) forests. It's been called "the Amazon of the North." The only northern refuge of the endangered whooping crane lies due north of Fort McMurray. Herds of wood buffalo and woodland caribou live back in the big woods. And wolves.

It was the boreal forest, but not the forest primeval. My road, Alberta Highway 881, had been built to open up the land south of Fort McMurray for oil sands development, which likely meant the movement of heavy equipment and a paved surface. In looking at the road online back home, I had noticed that there was no Google Street View imagery, and I couldn't confirm what surface I'd encounter. Experience told me that no Street View often meant a dirt road. But until I got to Fort McMurray I couldn't be sure. I had found no accounts on the internet of any cyclist who had ever ridden this road. Indeed, I wouldn't see another touring cyclist until southern Saskatchewan, almost 700 miles away. It would be a very lonely ride.

I wasn't excited to ride hundreds of miles of dirt. The evening before I left, I was nervous, pacing, thinking of the roads, the weather, and all the things that I might need that I wasn't bringing along. Stir-crazy in my room, I wandered down the industrial-plaid hallway to the front desk. The manager of the motel was alternately amused and curious about the idea of someone riding a bicycle out of town. I asked him if he'd ever been down 881.

"Sure, a buddy and I took a fishing trip down that road a few months back."

"So it's paved?"

"Yeah, and it's in pretty good shape. We didn't have any trouble. I've got a couple of pictures." He started scrolling through his phone. "Yeah, here she is. And her cubs too. Right on the side of the road. We got some pretty good shots, but we definitely stayed in the car."

I knew there were bears, but seeing them on his phone brought them a little closer. I thanked him and scurried back to my room for a string of laptop searches. What should I do if I saw a bear? Could I ride faster than a bear? I have a bright flashing taillight. Are bears attracted to strobes? Do they hate strobes? The internet, font of all knowledge, was less than helpful, though it did have some upbeat GoPro videos of bears chasing cyclists, from the rider's point of view. I settled into an uneasy night's sleep.

A raucous crowd of young roughnecks with beards and ball caps were in the breakfast room the next morning, before leaving to work in the oil patch. They laughed and tossed insults across the motel waffles. After they loaded onto their bus, I rolled the loaded bike through the double doors and into bright sun and heat. This first part of the ride to North Dakota was in many ways the hardest. I carried no camping equipment, which cut down dramatically on the weight I'd have to haul. But it meant that each day I needed to make it to a bed and a roof by nightfall. There was one more sign before leaving town, a giant, digital message board:

SHARE THE ROAD

I was encouraged for an instant. A big, bright, enlightened policy in Alberta. Then the sign changed:

WITH BIG RIGS

I certainly wouldn't be arguing with them.

I had done a fair bit of training before Alberta, but none with a fully loaded bike. I was carrying about thirty pounds without all the water. In all of those winter workout sessions at the gym, I had been wired into my own little world like everyone else, with earbuds and a playlist. Now as I paused by the roadside, I called up a line from a song I had played a hundred times: *Go forth and have no fear.*

Riding south out of town was not through the deep forest that I had imagined but rather miles of blackened, bare trees. The smell of ash was long gone, and below the trunks were the signs of rebirth. Willow and other scrub growth were coming up from what used to be the forest floor. The burn area extends 30 miles south of Fort McMurray, all along the road I was riding. An image of the fire at its peak, near my first destination, Anzac, shows the sky lit up to the north.

Riding out into the sticks of burned-out trees, I encountered my first problem with the fauna of northern Alberta, and it wasn't bears. Alberta horseflies could fly as fast as I could ride. I quickly acquired my own personal cloud (unfortunately not the online variety). They were looking for any non-moving, non-covered patch of skin. In fact, the overachievers were working their way through my thin riding jersey. I would have given anything to get rid of them. But there is, in fact, a god, and she has a wicked sense of humor. A stiff headwind came up and would be my companion for the rest of the day. The flies were gone, but my legs paid for it. I had to shift down into low gears to pull up even gentle hills. It was pushing 90°F, and I was rapidly blowing through water and Gatorade. Mostly it was very quiet, except when one of those big rigs was bearing down. Then I would instinctively tense up and mind the white line as tires would scream from behind.

Farther from Fort McMurray, traces of civilization became rare. A sign or even a roadside reflector became cause for minor celebration, if only for a place to lean the bike. Next to one grove of burned trees was a panel: FIRE PERMITS REQUIRED MARCH 1–OCTOBER 31. Looking more closely as I stopped for a water break, I realized that *March 1* was a patch, presumably covering some later date. Here was climate change deep in the forest. As with everywhere else in western North America, the fire season has expanded, leaving more chance for ignition during the course of the year. In the United States, climate change has led to fire seasons that are, on average, seventy-eight days longer now than they were in 1970.[1] For California, there is no off-season any more.[2]

Gradually the burn area fell behind, and the dark boreal forest folded in around the road. The dense spruce smelled like a walk through a Christmas tree lot, but for days on end. Occasionally in a roadside pond, a duck, startled by the commotion, would suddenly flap across the water.

My destination for the evening after a modest 41 miles of riding was the Surmont Lodge man camp. I'd seen other camps from the air on my flight—giant, prefabricated white structures built for hundreds if not thousands of workers. This camp was built to serve the Surmont project, a massive development of almost four hundred wells out in the forest. Unlike the open pit mines north of Fort McMurray, Surmont is referred to as an "in situ" project. Steam is injected into underground oil sand deposits to liquify the bitumen, oil that is too heavy to flow on its own.

The in-situ projects don't involve the physical removal of the oil sands, but the extraction process still requires a massive amount of energy. It can take 750 cubic feet of natural gas to heat the steam for a single barrel of the tens of thousands of barrels of oil pumped out per day.[3] These giant projects are typically out of sight from the road. But on the way to the Surmont Lodge, I passed white and silver storage tanks on a hilltop. Not far from there was a string of giant pipelines, three abreast, pumping carbon out of the ground, south to the States, down into our gas tanks, and up into the air.

Riding south in a steady drizzle on 881 brought an ominous sight: a huge abandoned man camp by the road. I didn't realize how exhausted I'd be on that first day, with the headwind and the weight on the bike. I had called ahead to Surmont to make a reservation long before, but if they had closed in the meantime, I was in trouble. At the Surmont sign, I turned onto a dirt road, the start of a winter ice road to Saskatchewan. Not far in were the telltale white prefab trailers and an empty, muddy parking lot behind a chain link fence.

The gate was open, but no one was at the guard shack. I tiptoed around the mud puddles to a doorway. The long entrance hall was lined with boot scrapers and boot racks, but just a few pairs of boots. Heather, a robust, friendly woman from Nova Scotia, came out of the office. We'd spoken on the phone.

"Sorry I didn't see you come in. We've got a pretty small crew these days. But I can fix you up with a real quiet room. You'll have the whole wing to yourself."

"It doesn't seem very busy. Not too much going on at the project?"

"Yeah, it's way down from the peak a few years back. Everybody's hoping it'll come back around with the oil prices. Have to say, we don't get many bicycles in here. You may be the first. So get yourself settled in and come down for dinner."

The bike cleats on my shoes echoed down the cavernous hallway, one of many wings to the facility. The walls were paneled with fake wood. Underfoot were the seams where the prefab sections had been bolted together. I squeezed the bike into the tiny room, only able to close the door after taking the front wheel off.

If dining-hall attendance was any measure, I was one of about half a dozen people staying at the camp that night. I was the beneficiary of a prix fixe buffet, with all the dinner (and breakfast and lunch) I could eat or carry away. Dinner was quintessential cafeteria food: bread, soup, ravioli, Salisbury steak, green beans, salad from a bar, pie, and soft ice cream from a dispenser. After a day of banging into big wind, I was ravenous and piled all that food onto my tray. A giant moose head presided over the hall. I settled in at a table. I asked one of the contractors at the next table about the sparse occupancy.

"It's not like it was a few years back. They've got production going, about 140,000 barrels a day, but nobody's drilling new wells. We're basically a maintenance crew."

I had noticed the armored, bear-safe trash cans outside. An after-dinner stroll wasn't that appealing. I walked back down into my own personal wing and stretched out in bed. The TV had a flickering connection to the outside world of the Edmonton stations. I drifted off, the only light amid hundreds of empty rooms. A loud infomercial at 2:00 A.M. startled me awake, utterly disoriented.

The crew had already left by the time I got up the next morning, and I had the dining hall to myself. I had wanted to learn more about life in the camp, but the quiet was actually the result of changes in the industry itself. When oil prices tanked earlier in the decade, the industry both cut back on new drilling and learned to do more with automation.

I motored through a stack of pancakes in solitude, stuffed my panniers full of lunch, and got ready to head back out on the road. Heather told me that there wasn't any place to stop from Surmont to the next night's destination, Conklin. That wasn't quite true, but after seeing how empty the road was, I could see why she thought that.

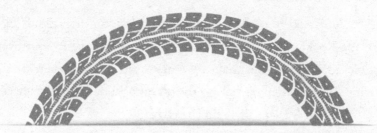

5

"SIGNIFICANT UNCERTAINTIES"

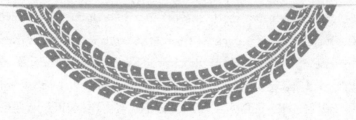

I leaned the bike up against the railing of the man camp and hung the panniers on the bike. "Uncertain" was a good adjective for the way I was feeling. But in an odd way I'd been here before. I'd experienced the oil industry from two very different perspectives: offshore Louisiana as a young drifter, and inside the Beltway as a middle-aged manager. The first one, on the Gulf oil rigs, was the more perilous of the two, at least as far as bodily risk goes.

In 1974, just out of college and without a job, I drove south from my home in Virginia, bound for the oil fields. It was a logical extension of my career ladder as a derelict. I had hitchhiked across the country several times, been a lumberjack on the Front Range of the Rockies, and spent three weeks backpacking in southwest Colorado. The reasons that called me to the offshore then are the same that call young people to the oil sands today. The money was good, and it was out there on the edge.

I rolled through New Orleans and onto the Delta of the Mississippi. The Delta was flat, with few buildings of any substance. Most were the sheet-metal structures of the oil contractors. Hurricanes flattened the Delta twice in the 1960s, and the towns showed scant evidence of rebuilding. I drove to the very end of the road: Venice, Louisiana.

My first job had me flying off in a helicopter for two weeks. The next time I was back on shore, the contractor's office directed me to a lonely dirt road at four in the morning. At the end of that dirt road, a crew boat bobbed under a solitary cone of light. I walked onto the gangway and looked down. The water was muddy, swirling. Black water. The muddy Mississippi.

The cabin was brightly lit and smelled of coffee. I blinked as my eyes adjusted. An overweight man with a clipboard looked up.

"What can I do for you?"

"I'm with Superior Welding. Headed for Battledore Reef," I said.

He squinted at his sheet and made a check mark.

"Have a seat. Underway in ten."

There wasn't much conversation that early. Much of the crew was seriously hungover from a weekend in the Big Easy. The crew boat engine rumbled to life and we chugged onto the river. A pack of cards came out and money flashed. I was just a kid, but I knew enough to stay out of their game. A novice would get picked clean. I took a seat by the window as we left the river for the Gulf of Mexico. Our job was miles offshore, out of sight of land, not all that far from where the Deepwater Horizon platform would blow years later. The oil rigs at first light glittered on the horizon.

The offshore wasn't so much a lawless place as a place to get away from the law. A fair number of the men on the rigs seemed to be on the run from something, so there wasn't a lot of trust going around. In the current era of instant web access to bank accounts, it's hard to imagine a strictly cash economy, but there we were. I brought along a copy of Shakespeare, which came in handy. I kept my cash in *The Merchant of Venice*.

The rigs weren't easy places to work. A film of oil perpetually coated the steel grating on the decks, so footing was dicey. We worked up on the high iron, looking down on sunfish gliding through the Gulf far below. Our company did a lot of welding, and the idea of sparks splashing around on a platform pumping crude didn't inspire much confidence. From pipes to I-beams to cables, the oil patch was a place where, then as now, big steel flew around.

One of our rigs was on the spiderweb of channels, or passes, of the Mississippi on its way through the Delta. On a buggy, sultry day, we were trying to connect a pipeline to the rig, pulling a hundred-yard pipe through a muddy field using a dragline, basically a giant crane. The cable from the dragline was attached to the front of the pipeline. I was working with a Cajun man from the crew, and our job was to guide the pipe through the field. With a short length of rope, the two of us were slogging through the mud, lifting up the front of the pipe to keep it moving through the tall grass. We were in sight of the dragline, but it was still the better part of half a mile away. The motor on the winch strained in the distance. I put down the rope to mop my brow.

I never knew what the front of the pipe hung up on. Somewhere down in the mud was something big and metal, and the pipe wasn't moving no matter how much we pulled. We waved frantically to the guy in the dragline cab, but he wasn't paying attention. The cable tightened up and began to sing. Then came a ping-ping-ping, the sound of cable strands popping. I turned to look at the Cajun, his eyes wide.

"Down!" he yelled.

I hit the ground to the sound of a crack like a gunshot. The cable snapped over our heads with enough force to cut through a torso. I watched as the other end of the cable danced in the sky for a moment, then settled back down to earth. I kept my cheek to the mud for long seconds, my breath in short, rapid puffs.

After one 106-hour week, the company cut our crew loose and shipped us back to the beach. The scuttlebutt was that another job might be coming, so they convinced us to stay in their bunkhouse in Venice, at least long enough to pick up our pay envelopes. The place had linoleum floors and a stale man-smell. One rainy night brought a call in a phone booth and the realization that my girl-friend and I would not be getting back together again. I slid the black plastic phone into the stainless-steel receiver and heard the coins drop into the box. I exhaled, and resolved that Venice, Louisiana, was as low as I would ever get. I drove out the next morning.

My second interaction with the oil industry wasn't face-in-the-mud up close but more like the gravitational pull of a distant planet. In 1998, twenty-three years after my time on the rigs, after ship driving, graduate school, and working for NOAA on shore, I was appointed to head the US Global Change Research Program Office. The position was less grand than it sounded. Our job was to coordinate climate research across federal agencies, a task not unlike herding cats. One of our jobs was to ensure support for US contributions to the Intergovernmental Panel on Climate Change (IPCC), the international body charged with evaluating the current state of climate science. The Panel was about eight years old then, nine years before it would receive the Nobel Peace Prize.

The basic science of climate was settled way back then as now. The planet is getting warmer, steadily. Sea level is rising, steadily. From Alberta to California to Siberia, forests dry out and burn far more readily now than several decades ago. Heat waves and big storms are more intense, largely driven by climate change. Our climate is on a path to warm beyond the range of what has been experienced over the past millions of years. What's driving climate change is also clear: the burning of fossil fuels—oil, coal, and natural gas—with a lesser contribution from deforestation.

At the Global Change office, I worked with an owl-eyed Texan by the name of Rick Piltz. Gray-haired, looming, soft-spoken, Rick always seemed to wear a little scowl that would readily break into a knowing smile on recounting any of the dozens of little intrigues that swirled around Washington. He would proceed to delve into the worst-case scenarios, his darkest projections of what might happen. His tagline was always: "Nothing to keep us from having a good time on the weekends."

Though on weekends, Rick was something less than a party animal. As befits a former professor, he read incessantly, always aware of the arc of history. His wife, Karen Metchis, described their weekend routine:

> My daughter and I would go out shopping, leaving him in his easy chair reading a book or wandering around the Internet. We'd come back a couple of hours later to find him in exactly the same place. He loved getting lost in the world of ideas.

One of Rick's jobs was to shepherd through an annual report of agency activities and findings on climate titled *Our Changing Planet*. The report was then cleared through a White House office. It was all pretty tepid, bureaucratic stuff.

I left the Global Change office in 2000 but kept in touch with Rick. Not only were we friends, but I coordinated the NOAA contributions to *Our Changing Planet*. The George W. Bush Administration took office in 2001. Not long after, a confidential report to the White House by Republican political consultant Frank Luntz on environmental communication emerged in the pages of the *New York Times*. Under the heading "Winning the Global Warming Debate," Luntz wrote:

> Should the public come to believe that the scientific issues are settled, their views about global warming

will change accordingly. *Therefore, you need to continue to make the lack of scientific certainty a primary issue in the debate . . . the scientific debate is closing (against us) but not yet closed. There is still a window of opportunity to challenge the science.* [Emphasis his][1]

We didn't realize it at the time, but this was the Rosetta Stone, the playbook for what was to happen. This defined how climate change would be framed during the Bush Administration and for years after. Starting in 2002, peculiar edits from the White House began to appear in *Our Changing Planet.* "Uncertainty" in scientific findings became "significant and fundamental uncertainty." "Earth is undergoing relatively rapid change" became "Earth may be undergoing relatively rapid change." Paragraphs describing how warming would cause reductions in mountain glaciers were struck as "speculative musings." Many of those mountain glaciers have vanished today.

The author of the edits was Philip Cooney, chief of staff at the Orwellian-sounding White House Council on Environmental Quality. He had come there from the American Petroleum Institute (API), the Washington lobbying organization for the oil and gas industry, which had worked for years opposing limits on fossil fuel emissions. Cooney, an economist and lawyer with no science background, had been a "climate change leader" at API.

I saw the edits. Organizations funded by oil companies and dedicated to obfuscation on climate had been around for a long time. This was different. Now the fog machine was part of the government. From a distance, like the pull of gravity, the oil industry was working to bend the shape of the science. Concetta says that I didn't usually take the office home, but she remembers me walking through the door with pursed lips the day that the most extentive edits came through.

I was upset. Rick was more than upset. In 2005, he had been wrestling up close with the White House's warping of the science for years, and it was tearing him up. After long conversations with Karen, she told him, "You decide. Whatever you need to do, just do, and I'll support that."

Rick bet it all. He later said, "I resigned without a plan except that I'm now free to speak."[2]

He parked his car one night on Pennsylvania Avenue, in sight of the White House, and rolled a handcart full of boxes out the front door of the office. Shortly after, Rick Piltz sent a FedEx package to the *New York Times* with the collected, edited drafts of *Our Changing Planet* from the previous two years.[3] At the White House, the *Times* reporter asked if Mr. Cooney might have some comment.

"We don't put Phil Cooney on the record," the White House representative said. "He's not a cleared spokesman." On June 8, 2005, the story of the softening of climate language appeared on the front page of the *Times*. Two days later, Cooney resigned to take a new position at ExxonMobil in Dallas.

Rick had less of a soft landing. As a contractor with the Global Change office, he had no whistleblower protections. He encountered Stephen Schneider, one of the leading climate scientists, at a science meeting. As Rick recounted, he came up during a break and said, "Wow, I heard that you fell on your sword."

"Hey, I'm just saying what you guys are saying," Rick replied.

"What were you thinking? I'm tenured at Stanford. I can do anything! You're on the street!"

"I'm better off on the street," Rick said. "This is good. Look what I can do here."

After resigning, Rick spent the next nine months without income or benefits. He cashed in his retirement money and refinanced his home loan to found a government monitoring site called "Climate Science Watch." Two years later, he testified about the political interference with climate science before the House Committee on Oversight and Government Reform.

Yet it didn't turn out so badly for Rick and Karen. Before he pushed the button on his resignation email, she told him, "We'll be fine." With the benefit of her steady job and some belt-tightening, they got through. After resigning, he felt like a great weight had been lifted. He was awarded the Ridenhour Prize for Truth-Telling in 2006. Today Rick's picture hangs in the conference room of the Government Accountability Project, a private nonprofit, which sponsored him until his passing in 2014.

The "significant uncertainty" messaging, promoted by the Luntz report and implemented by the White House, has had a lasting legacy. While a majority of Americans today understand that global warming is mostly human-caused, only one in six understand how strong the level of consensus among scientists is—namely that more than 90 percent of climate scientists think human-caused global warming is happening.[4] In November 2019, more that 11,000 scientists from 153 countries signed a report warning that the planet "clearly and unequivocally faces a climate emergency."[5] Yet what Rick Piltz revealed continues nearly twenty years later. Climate reports coming out of the Department of the Interior through 2019 routinely contain language inserted by the front office stating that there is a lack of consensus among scientists that the earth is warming.[6]

Curiously, Frank Luntz is now an enthusiastic proponent of action on climate change. In 2019, he testified before Congress on the best messaging strategy for climate change. His change of heart happened sometime after he was awakened in his California home at 3:00 A.M. to an emergency evacuation order. His home survived the 2017 California firestorms, but many of his neighbors' homes did not.[7]

Ken Caldeira, a climate scientist at the Carnegie Institution for Science at Stanford, has a habit of asking new graduate students to name the largest fundamental breakthrough in climate physics since 1979. It's a trick question. There has been no fundamental breakthrough.[8]

That's not to say there haven't been major advances in the many different fields of climate science. But I think about some imaginary tribunal held in the future, when the full impacts of the warming planet have become apparent. The inquisitor might ask the question about human-induced climate change: What did we know and when did we know it? The answer is that we've known the basics for about forty years. The fog machine has been working for a long time.

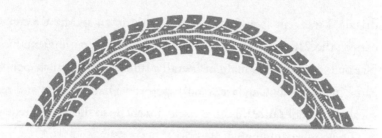

6

THE MAGIC JUICE BOX

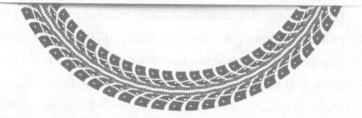

R eal fog enveloped me as I pushed the food-laden bike out through the muddy parking lot of Surmont Lodge and back onto the road. This was a quiet day in the dense birch forest, a little longer ride than the day before but without the killer headwind. I thought that I'd have nowhere to stop for the whole day until a little flash of white caught my eye.

The 218 Trading Post, a convenience store named for the kilometer marker on Route 881, was barely visible from the road. A hand-painted plywood sign bolted to the side of a bare white shipping container announced its presence. After a couple of hours in drizzle, I was only looking for a place to sit down and have a cup of coffee. The 218 is on the land of the Chipewyan Prairie Dene First Nation, and a steady stream of friendly, funny, First Nations people came by my table. Candace, a tall, sad-eyed Dene in jeans and red sneakers, presided over this little world. I went up to the counter to pay.

"Don't worry about it," she said.

"You sure?"

"No problem." She paused to talk, in that way that people who've been isolated sometimes open up to strangers. "Yeah, those guys were definitely high."

"Really?"

"Oh, I can tell. It's easy. Been there. Walked in the dark for a long time. It tears you up, tears your family up. Opioids are all over this place. But I've been sober for a solid two years now."

"That must have been really hard. How'd you do it?"

"We've got a different way. Out in the woods, healing camps. I even helped run one. Help you get connected to the land again, get away from all the booze, the drugs, the TV. It helps to get back to the way things were, to talk out all the bad stuff that's happened. The men even had a sweat lodge."

I could have stayed longer in the warm cocoon of the 218, but I needed to put some miles under my wheels. The rain had stopped as Candace let me take her picture. She leaned against the corrugated-steel wall of the shipping container. Her thumbs were hooked into her belt loops, her face not quite a smile, narrowed eyes looking at the camera but somehow focused on another place.

I wandered around the corner into a group of Dene talking outside. The bike was always a conversation starter. After I mentioned that I was from Washington, an older Dene man, his face crinkled into a smile, leaned over from behind the wheel of a pickup.

"Why don't you stop by Janvier tonight? I could set you up with some women."

"That's very kind, but I don't think my wife would approve."

"Aw, she's probably all painted up and walkin' up and down the Mall tonight."

I grinned. "Now you've got me suspicious. I'll have to check that out."

The pickup drove out through the mud puddles, and Candace came out to offer a piece of advice.

"I think it's just as well you moved on," she said in a low voice. "Janvier can be a rough place."

I left my warm spot at the 218 and got back out into a birch forest, white trunks silhouetted against leaves quaking in the breeze. After a quiet morning, the road was bustling with activity. I passed a Royal Canadian Mounted Police truck with blue lights flashing behind a white van. Not far past, two women walking along the road flagged me down. One was obviously flustered.

"We're from University of Alberta, working on a site out on one of the ponds. Our van was broken into. They got my phone, my laptop, a jacket, everything. The police said stuff may have been thrown away, and they said we should look by the side of the road. Could you look and call on my friend's phone if you see anything?"

I took the cell number and tried to focus on the roadside for the next thirty miles, without coming across anything. I thought of Candace's warning.

The evening brought a Ramada in the town of Conklin, downright luxurious in comparison to the man camp. Sitting on the check-in counter was a sign:

> DEAR GUESTS,
> PLEASE BEWARE OF BEARS. WE STRONGLY RECOMMEND YOU LOOK AROUND FIRST BEFORE HEADING OUT, EVEN JUST ON YOUR WAY TO THE VEHICLE.

I asked the front desk attendant about it. He shook his head.

"Yeah, whatever we say doesn't seem to matter. We had a couple of drunks chasing a bear around the parking lot last night."

By Conklin, I was getting into a familiar rhythm with the bike on the road. It became more of an extension of my body than an accessory. Spokes were like tendons, and I'd swear that I could fix that knee pain with a little chain lube. In the evenings, I kept promises to the bike in the motel room, tightening bolts and chasing down squeaks and chirps. My wardrobe was a bit limited, so the motel also brought the incessant march of laundry. Binder clips held a parade of wet clothes on curtains above the A/C vent.

Back on the road the next day I thought about the sign on the front desk, but bears were the least of my problems. All winter I had worked on preparing a long, detailed spreadsheet with each day's ride, mileage, and destination. But as Mike Tyson said, everybody has a plan until they get punched in the face.

The next day was to be the longest, emptiest stretch of the ride: seventy-five miles to Lac La Biche with no towns, stores, or even buildings. That morning, the cheery Edmonton weatherman talked about a beautiful day "except for 30- to 50-kilometer-per-hour winds." I rolled out of Conklin at first light on a cold, raw morning. I was nearly 10 miles down the road when the rain started. It stung my face as I crawled up each hill. Hard pulls are nothing new, but my legs started screaming early that morning. Feeling like I was making no headway into the wind, each turn of the crank was an effort.

I was warm enough pushing the pedals, but got cold almost immediately once I took a break. I had an image of being halfway into the ride, borderline hypothermic, with no place to stop. So for the first time in almost twenty years of distance riding, I turned around because of the weather. I did have a certain satisfaction that afternoon in watching the trees sway in the rain from the comfort of my room.

Backtracking—the notion of giving up miles—was excruciating. Every one of those dozen hills would have to be mastered once more. The bus back to Fort McMurray was sitting in the Conklin parking lot when I returned, mocking my retreat. I wasn't completely committed yet. I could still bail on the ride.

Turning points are rarely dramatic. It's often the simple process of getting beaten, getting overwhelmed, then getting up again the next morning. I was in this mess, and I needed to ride out of it. I took another shot at the Conklin road after a night's rest, this time with a little more success. Low, ragged clouds swept across the deep green landscape. Bursts of rain came through. I'd get wet, grudgingly put on my rain gear, and that would make the rain stop.

The Doppler showed a pinwheel of rain sitting over this part of Alberta. In the distance, I could see I was riding toward a little patch of blue sky. It was there all day, like the rabbit hung in front of greyhounds at the dog track. I never caught up with it.

With the exception of the truck traffic, this was one of the least populated stretches of road I'd ever ridden. As days wore on, I began to notice my body changing as well as the land. One of the marks of coming into condition is the ability to generate second and third and fourth winds out on the road. This became the rhythm of the days. Sinews began to stand out on my legs. Despite continued use of sunscreen, my skin was darkening. One dinner, I noticed a dark spot on the back of my hand that wouldn't wash off. Was this some exotic fungus that I'd picked up? Then I noticed the same spot on the other hand. It wasn't until I got back to the room that I realized that my riding gloves had a circular hole in the back, its imprint now tattooed on my skin.

When I lay in bed at night, I could feel my body rebuilding. The strong pulse in my legs would keep me awake, but not for too long. A dark-black sleep would come with bright, vivid dreams, just out of reach on awakening, as though my mind were compensating for the solitude of the miles of deep forest.

A warm, sweet bed-and-breakfast awaited at Lac La Biche. My host, John, had the build of a former hockey player. He and his wife, Denise, invited me to their dinner table that night.

"I sure went through some burn area south of Fort McMurray," I said.

"We saw our own piece of that," said John. "Not the fire, but back in 2016, we had families pouring out of Fort Mac. Everybody was opening their homes to evacuees. We had a full house here."

"That's a long stretch between here and there," I said. "Some of my days are crazy long because I want to avoid carrying camping gear. Maybe I should have brought it along. I know that bears can be a problem for camping out, but I could have hung my food bag in a tree."

John rubbed his chin. "I'd worry more about the wolves." *Real wolves?* I thought. "Yeah, I remember one night I was backing up my truck up there and saw a couple of eyes in the rear view. When I shone a light, there was a big old wolf."

"Well then. I guess I'll stay inside." There's a qualitative difference between hanging food and being food.

The next day was another long one on the bike. But the morning delivered a break: a bright blue sky washed clean, a few openings in the forest, a few farm clearings. I rolled through the settlement of the Kikino Metis, people who are of mixed aboriginal and European ancestry.

Thunderheads began to build from the west as I started into fifteen miles of steady climbing. I tried to outrun a particularly dark cloud. I kept looking over my shoulder, watching the ragged squall line close on me. In minutes, it was crackling right over my head, lightning rippling through the sky. The bike was the only metal around, so I got away from it and sat down low in the rough stubble of the roadside grass.

Looking into the distance, the forest seemed to be out of focus. Rather, it was blurred behind the gray edge of the rain front moving ominously up the road. I zipped my jacket up to my chin and braced as the sheets of water swept over me. Drenched again, I began to shake. The rain started to bounce. I pulled my knees into my chest to stay warm, shivering, watching as ice pellets gathered in the folds of my wind pants. Then, a different noise.

I turned my head to the sound of crunching gravel. In the middle of the tempest, an SUV pulled up and its window rolled down. A crust of ice remained where the bottom of the window had been. The driver, a young mom, leaned over her elementary-school-age daughter in the front seat.

"Do you want a place to warm up?" she shouted.

"I'm pretty wet," I yelled between the lightning flashes.

"That's fine. Come on in." She turned to the girl. "Okay, into the back seat. Quick now."

The girl squeezed between the seats, not pleased to be shoe-horned between her two little brothers. Outside, I shook off like a spaniel, then opened the door and got in. I took a first good look at my savior. She had short blonde hair and tiny stud earrings, too small for an infant to grab. She was in crowd-control mode.

"Stop kicking the seat," she growled at her daughter.

"That's 'cause he was kicking me," was the response.

The mom shot a look, then turned to me. "Are you the cyclist who rode down from Fort McMurray?"

"How did you know?"

"There's a grapevine out here. A friend at the Settlement told me about some guy riding through who didn't look local. I was driving right through this mess when I saw you. I'm Jesse."

She offered me coffee from her thermos. "I'm very picky about my coffee. Cream and just the tiniest bit of sugar. Hope it's okay."

It was the best I ever had.

"So, how'd you end up out here?" she asked.

"Heard a lot about the oil development here from down in the States where I live," I said. "Thought I'd come see it firsthand. And I ride my bike a lot."

"Papa was a bus driver up at Fort McMurray, so we grew up around it. The land is sick up there. We've done protest walks around the tar sands. Sometimes they draw so much water out of the river that you can't even get a canoe in." She paused to look as the storm clouds began to show breaks.

"Lot of changes around here too. I remember going and jumping in our lake and it would be clear and you could see the bottom and see fish swimming. It's a mess now. I'm afraid my kids are going to think this is all normal."

I mumbled something incoherent, but I was mostly focused on the wonderful stream of warmth coming out of the heater vent. The kids were getting restless—"past their nap time," Jesse said. I could see that she was looking around, trying to give me something to help me on my way. Then she reached down and gave me just what you'd expect from a mom with a van full of kids: a juice box. The sky was clearing as they drove away.

I'd like to say the rest of the day went fine, but it wasn't like that. Hours of riding remained. The chain started making hamster-wheel noises. The downpour had washed all the oil out, so I had to stop and re-oil. Not long after, a flat had me back sitting by the roadside. I screamed every bad word that I knew and listened as they echoed through the forest.

All my planning pointed to Vilna as the first rooms I could find on the way south. But I'd had trouble reaching the hotel on the phone. As I rolled into town, I saw the reason: It was closed, and with it my dreams of a Hemingway-esque night at the Roaring Thunder Saloon. A man with a gray beard working on a ladder called to me.

"It shut down a couple months ago," he said. "It's been a hard year."

Another nine miles down the road in Spedden, I pulled into the Northern Lights Truck Stop and Fas Gas at twilight. Strung out to the side of the convenience store were the first motel rooms for seventy miles. Utterly exhausted, I went to pay at the counter. Three people were behind me as I suddenly realized that I couldn't find my wallet in the bag I had brought in. I desperately pawed through the bag, mumbling, "I'm sorry . . . I'm having trouble . . . I have to go out and look on the bike . . ."

The attendant gave me a condescending look. "Is that it?" She pointed to my wallet, on the shelf at eye level, where I had put it.

In the room, after a shower, I had wet clothes hanging to dry from every possible hook and corner and angle in the room. The ubiquitous convenience store sandwich was warming in the microwave. It was a lonely place, and the road had taken a lot out of me. Though I hardly realized it, at the Northern Lights I was on the border, the southern frontier of the boreal forest, ready to break out onto the prairie. Many hundreds of miles remained, but none so remote as what I had just passed through. I feasted on the cheap sandwich and reveled in the shelter. And on the desk sat a juice box to remind me that the afternoon's act of grace actually happened.

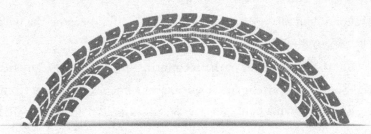

7

THE WATER PROTECTORS

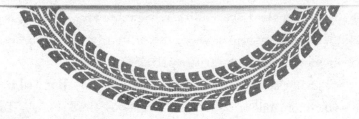

Meeting Jesse and riding through the Kikino Metis reminded me that any account of oil in the north is incomplete without the stories of Indigenous people. Many of the hardest battles in the development of oil in the North—both in Canada and the United States—have been fought by the people who first lived here. The driving motivation behind the battles have been far more about water than climate. The most celebrated recent pipeline battle in the United States was the routing of the Dakota Access Pipeline, just upstream from the Standing Rock Reservation in North

Dakota. *Mni wiconi*, Lakota for "water is life," became the battle cry of Standing Rock.

But there's a danger in the recounting of an old white guy from the South. I know climate science, and I know what I saw from a bicycle seat. Yet we all have our blind spots. An encounter thousands of miles to the east, in Ontario, just across the border from Buffalo, New York, highlighted the issue for me. Lezlie Harper Wells, a descendant of the enslaved people who escaped to Canada in the nineteenth century via the Underground Railroad, took Concetta and me on a historical tour. She included a stop at a recently completed memorial to these black immigrants in the town of Niagara-on-the-Lake. A little reluctant to take us through, she pointed out a few places where the primarily white sponsors of the memorial didn't get the story straight.

"It's good that they did this, I guess," she said. "But it's kind of like when someone offers to paint your house, and you say, 'That's great!' Then you find out they painted it purple."

To help me avoid painting a purple house, the focus here is on three Indigenous voices: Eriel Tchekwie Deranger from Alberta; Nick Estes from North Dakota; and Winona LaDuke from Minnesota. Theirs are stories of three different pipeline battles, on both sides of the border. Their voices portray Indigenous people not as victims of catastrophes in a climate story—though there has been trauma to be sure—but as knowledge keepers, land defenders, and water protectors. In short, as people from whom we can learn.

Eriel Tchekwie Deranger is a young member of the Athabasca Chipewyan First Nation and the cofounder and director of Indigenous Climate Action, a nonprofit organization based in Edmonton. She has intense eyes behind dark-framed glasses. In interviews, she speaks of when her father, a trapper, would take her on his trap lines years ago. Today those places are where the giant dump trucks roll across what is arguably the world's largest industrial development. For five years, Deranger led protesters on the roads though the mines. The protests were known as Tar Sands Healing Walks, the same ones that Jesse had told me about.

Lands of the Water Protectors, bicycle route in dashed line.

Growth of the oil sands mines, with their excavators and lines of trucks, has been largely limited by the ability to get the oil out of Alberta. If pipelines don't expand, the mines won't either. The three major proposed or expanded pipelines are the Keystone XL, Trans Mountain, and Enbridge Line 3. Keystone is well known in the United States. The Trans Mountain is a proposed expansion of an existing line from Alberta through British Columbia, and it has roiled Canadian politics in much the same way Keystone has in the States. When the US corporation Kinder Morgan backed out of the Trans Mountain expansion, Canadian Prime Minister Justin Trudeau announced that the government would take the extraordinary step of purchasing the pipeline for $4.5 billion Canadian dollars to ensure its construction.

Deranger is one of the prominent voices of opposition. In one interview, she stated, "What we're seeing now is that a government has gone from supporting fossil fuel companies that violate Indigenous rights to actually becoming one themselves."[1]

By no means is there unanimous opposition by Indigenous people to oil sands development. Some First Nations groups are exploring the possibility of partial ownership of the Trans Mountain pipeline.[2] Chief Allan Adam of the Athabasca Chipewyan First Nation, Deranger's people, once led celebrities like Darryl Hannah and Neil Young on tours in opposition to the oil sands. Bryan, the engineer who had driven me through the oil sands, certainly remembered Daryl Hannah's visit from LA. Yet late in 2018, Chief Adam signed on in support of the largest oil sands project ever proposed, Teck Resources' Frontier Project—to be built in his community's own backyard.

How did Chief Adam's change of heart come to pass? After so many battles for so long, the view of the Athabasca Chipewyan leadership began to change. The question was whether to keep battling the oil patch when there was no reason to believe that the outcome would be any different this time. Chief Adam summarized it:

> The sad scenario is that I would have loved to fight, and I still love to fight today, but there has to be a time when you have to draw the line . . . [The Frontier Project] is bringing in a lot of funds for our First Nation, where we can at least develop our infrastructures, get our people well educated, hopefully do more business.[3]

I wonder, if I were in his situation, whether I might have come to the same conclusion. Deranger, his longtime colleague, described the conflict this way:

> My community, like everyone else in the province, has been forced into an economic relationship with an industry that asks us to compromise and sacrifice our lands, waterways, culture and rights so that our people can put food on the table and keep a roof over their heads. The other day I was asked if I felt betrayed by Chief Allan Adam's seemingly sudden change of heart on issues in the Alberta tarsands. . . . My answer was a resounding no. I said no because I didn't want to be baited into fighting my own people. This is bigger

than what Chief Adam has just said or done. This is a symptom of the neo-colonial agenda. My community, just like the other Cree, Dene and Métis communities that have stepped up in support of this atrocious industry, have been forced into a corner through years of concerted pressure by oil and gas companies in collusion with government to accept the tar sands as our fate.[4]

Chief Adam was forced into a choice with no good outcomes. He could accept the poverty of many of the northern First Nation settlements or the health threats and devastation of the land that comes with oil development. It reminded me of another place I've ridden, thousands of miles away. The banks of the Mississippi River from New Orleans north to Baton Rouge are lined with refineries and petrochemical plants. Nearby are predominantly African American communities. Many of their residents are descended from enslaved people who worked the sugar plantations that were once along the river. These days the area that stretches along the river has another nickname: Cancer Alley.

It occurred to me that I'd flown over one image for what the future might hold for the Athabasca Chipewyan. The town of Fort McMurray, where I'd stayed, is at some distance from the worst impacts of the oil sands mines, with the closest mines about 7 miles (11 kilometers) away. Buses roll north every day from the town to the mines. Fort McMurray is largely white. Fort McKay is the town that the pilot pointed out to me when he flew me over the oil sands, the town where breezes from nearby processing plants smelling like cat pee and rotten eggs waft over from time to time. It's probably not coincidental that it's a First Nations community.

In a strictly economic sense Fort McKay has profited from the adjacent oil industry.[5] Oil-field service firms there are largely Indigenous-owned, and average incomes exceed those of Alberta as a whole. But residents also report elevated rates of rare cancers and worry about an apparent increase in the number of premature births. They must drink bottled water, can no longer eat fish from the Athabasca River, and have been warned to take short, luke-warm showers to avoid skin problems.[6]

Deranger sees the issue of development in oil sands as more than just one battle or protecting one piece of land. She has been a leader on incorporating Indigenous rights into the climate and environmental movements. "It's more than the right to be con-sulted. It's the right to say no," she says. "Globally, Indigenous people are becoming the face of the environmental movement."[7]

The protests, both Canadian and global, may be having an effect. In February 2020, Teck Resources withdrew its application for the massive Frontier Mine, citing the government's failure to reconcile resource development and climate change.[8] Low oil prices and lack of pipeline capacity to get the oil out of Alberta certainly played a role. Global investors were not prepared to take the kind of heat that investments in the oil sands would bring. And the crash in oil prices following the coronavirus outbreak made it unlikely that they would revisit that decision.

A dusty, old button sits on my dresser proclaiming, "Rise With Standing Rock." The button came from a demonstration in Washington, DC, in March 2017. At that time the names Standing Rock and the Dakota Access Pipeline (DAPL) were in headlines around the world. The media spotlight came to focus on a windswept region of the North Dakota prairie on the banks of the Missouri River, just north of the Standing Rock Reservation. A pipeline carrying oil from the Bakken field was to cross the Missouri just north of the reservation and through the sacred lands of the Lakota. A great camp in the shape of a buffalo horn grew at this place in protest to the DAPL. This was the encampment of the Oceti Sakowin, or Seven Council Fires, the seven divisions of the Lakota, Dakota, and Nakota people, known as the Great Sioux Nation. The pipeline builders were up against formidable opposition.

Nick Estes is a prominent chronicler of the Standing Rock story with his book *Our History is the Future*.[9] Estes is a Lower Brule Lakota who moved twenty-three times growing up and remembers patching holes in the family car's floorboards with cardboard to keep the snow out.[10] He also read compulsively. These days he's a professor at the University of New Mexico.

Estes joined the Oceti Sakowin encampment. The source of the immediate opposition to the pipeline was water. The Missouri River is the water supply for the Standing Rock Reservation, and sooner or later, pipelines break. The touchstone for the movement became the Lakota phrase *mni wiconi*, or "water is life." Yet as Estes would say, this is not about one place or one pipeline. The fight against DAPL can only be understood in context, in the difficult history of Indigenous people in the Dakotas.

Estes's own home town of Chamberlain, North Dakota, began in the nineteenth century as the trade hub of Fort Kiowa. The river trading forts were hard places, especially for Indigenous women. Estes calls them "the first man camps."

Another gathering of the Great Sioux Nation and other Indigenous peoples took place 130 years ago but remains close in the Lakota memory. In 1890, in the approaching winter, the same time of year as the pipeline protests, a religious movement known as the Ghost Dance took place on the Pine Ridge Reservation in South Dakota. Participants in the Ghost Dance believed that, by performing the ritual dance, they would make lost ancestors reappear, the buffalo return, and nonbelievers vanish from the continent. They gathered in a natural fortress known as Stronghold Table at the edge of what is now Badlands National Park. I've camped at the entrance to the Stronghold. In the twisted, stunningly beautiful landscape of painted mesas and deep ravines, it's not hard to imagine the earth opening up.

In December of 1890, a group of Lakota bound for the Stronghold was intercepted and surrounded by the US Army at Wounded Knee Creek. In one of the darkest moments in US history, as many as three hundred Lakota were massacred. For the Lakota, for the Indigenous of North America, Wounded Knee is always in the background, part of what Estes calls "the Indian War that never ends."[11]

In the more recent context, the construction of giant dams along the Missouri River in the mid-twentieth century resulted in the flooding of tribal bottom lands and the relocation of more than a thousand Indigenous families in Montana and North and

South Dakota. This program, known as the Pick-Sloan Plan, destroyed more Indigenous land than any other public works project in US history, affecting twenty-three different reservation communities. However, the Army Corps of Engineers modified plans for the Garrison Dam in 1946 to protect the majority-white town of Williston, North Dakota.[12] According to Estes, the memory of the flooding was still fresh in the Standing Rock camps.

A half-century later, a similar dynamic was at play for the DAPL. The original route for the pipeline was to pass upstream of Bismarck, the predominantly white state capital. In part because of a perceived threat to the city's water supply, the pipeline route was moved to just north of Standing Rock.[13] As in Fort McKay, oil development seemed to happen close to Indigenous communities. The pipeline developers and state authorities must have thought it would attract less protest.

It didn't go that way. Throughout the summer of 2016, demonstrations grew at the site of the pipeline's proposed crossing of the Missouri. Protesters from the Sioux and other Indigenous nations, together with non-Indigenous activists, gathered at the Oceti Sakowin camp. On August 19, North Dakota Governor Jack Dalrymple issued a declaration of emergency, asking for assistance from the federal government, DAPL, and "any entity we can think of."[14] Police and private security from across the Plains gathered.

As fall gave way to the early winter, an amazing scene played out at the encampment. Anishinaabe author Louise Erdrich describes coming upon the camp in the sparsely populated country:

The presence of so many people catches at the heart. Snow-dusted tepees, neon pup tents, dark-olive military tents, brightly painted metal campers, and round solid yurts shelter hundreds on the floodplain where the Cannonball River meets the Missouri. Flags of Native Nations whip in the cutting wind, each speaking of solidarity with the Standing Rock tribe's opposition to the Dakota Access Pipeline.[15]

At the Oceti Sakowin camp, the Seven Nations rejoined. The camp was surrounded by other Indigenous national camps—over ninety Native nations—and by camps for non-Indigenous supporters. For Estes and many others, the rudimentary conditions were idyllic: a nation reunited, a dream come true. The protestors stayed on message, advancing through prayer and ceremony, staying nonviolent against the well-armed force beyond the fence.

As the call went out across the nation for support for Standing Rock, so did the call for law enforcement. Seventy-six jurisdictions responded to Dalrymple's call. Razor wire walled off the camps and drones circled overhead. Early one morning, earthmovers came, escorted by private security with dogs. Many protesters were bitten and/or arrested. In late November, police employed a new tactic, spraying a mist of water in the sub-freezing temperatures. An iconic *New Yorker* photo shows a Water Protector covered in ice kneeling in front of the razor wire and an armored Humvee.

There were moments when victory seemed within reach. In early December, groups of veterans from all over the country began to arrive at Standing Rock. On December 4, the Army Corps made a stunning announcement: It had denied the easement for the pipeline to cross under the Missouri River. Many protestors stayed, suspecting that the decision could be overturned by the incoming administration.

Pressure from the government, together with the brutal North Dakota winter, ultimately forced the abandonment of the Oceti Sakowin camp. The blockade of Standing Rock effectively shut down the casinos, the Reservation's primary source of income. Support of the giant camp sapped their meager resources. North Dakota issued an evacuation order in February, with arrests of the people remaining. Bulldozers cleared the camp.

Following his inauguration, President Trump reversed the Army Corps decision and allowed the easement. The DAPL was completed and Bakken oil flowed, to the approval of financial markets.[16] Meanwhile, leaks on other Dakota pipelines occur with a certain regularity: 407,000 gallons on the existing Keystone pipeline in 2017; 84,000 in North Dakota in 2019; and another 380,000 on the existing Keystone North Dakota Pipeline in 2019.[17]

On a snowy March day in 2017, I put on the "Rise with Standing Rock" button. Marching down Pennsylvania Avenue in Washington, I heard a chant from the Water Protectors: "We're cold. We're wet. We ain't done yet."

Standing Rock brought another Water Protector to the fore. Winona LaDuke, internationally known Ojibwe activist and two-time Green Party vice presidential candidate, joined the Oceti Sakowin. With long black hair and beaded necklaces, she is a grandmother and a storyteller, a woman who has lived much of her life on the road.

I wasn't the only one to have ridden along oil routes by alternate means. In 2016, with forty other Native Americans, LaDuke rode three days on horseback covering over 200 miles from Standing Rock to Tioga, in the Bakken. Tioga is where the Amerada Petroleum Corporation drilled the first North Dakota oil well a half-century earlier. The travelers called their journey "Ride Against the Current of Oil."

LaDuke's people have their own pipeline battle back in their home in Minnesota, and I would see the traces of it. As I crossed southern Saskatchewan, I came upon giant piles of pipe stacked in a field. This was one of the stockpiles for Enbridge Line 3. As with the Keystone XL and the Trans Mountain pipelines, these are the means by which oil sands bitumen is to be brought to international markets—"to tidewater," as they say in Alberta. In the case of Line 3, tidewater means Lake Superior. The proposed route of the pipeline skirts to the north and east of LaDuke's home, the White Earth Reservation of the Anishinaabe. Line 3 would also pass through the headwaters of the Mississippi River.

In LaDuke and her organization Honor the Earth, the pipeline company Enbridge discovered a most challenging opponent. The idea of a spill in their country, where the Anishinaabe harvest wild rice from some of Minnesota's 10,000 Lakes, was anathema. Along with lawsuits, court battles, and testimony before utility commissions and the state legislature, LaDuke sponsors and participates in a horseback "Ride of the Water Protectors," now in its seventh year. But she's after far more than stopping one or two pipelines. In a recent lecture, she states:

> We're at a moment in time when we have the ability to stop the prevailing white-men mentality from combusting the planet to oblivion. . . . We have a shot at keeping them from topping off any more mountains in order to provide coal to sell, not even here but to India. We have a shot at stopping some pretty extreme behavior at the end of the cannibal or *Wiindigoo* economy. This, in my estimation, is a great spiritual opportunity for us all. [18]

She's not alone. LaDuke speaks of another Indigenous leader from Wisconsin, Mike Wiggins Jr., who had been in a successful fight to stop a huge taconite mine there. He mentioned to her that he'd been at the Wisconsin state legislature, and he then said, "You know, seems like those people don't want to hang around another thousand years."

"Hanging around" in Wiggins's lingo translates to "sustainability" in mine. LaDuke speaks often of a prophecy known as the Time of the Seventh Fire. It is said that long ago, prophets came to the Anishinaabe people, and they said that the people would have a choice between two paths, one well-worn and scorched and the other green. She believes that the Time of the Seventh Fire is now.

As a scientist, my first instinct is to nod and mentally file away the story under the category of myths and superstitions. But in the foretellings of my community, the climate scientists, projections of the most powerful supercomputers of many nations go to some rather dark and apocalyptic places. When climate models described in the IPCC are extended out two hundred to a thousand years—the timeframes of Winona LaDuke—the global temperature increase is as much as 18°F.[19] That's enough to make large parts of the planet uninhabitable. Or, as the prophets might put it, scorched.

The prophecies of the Ojibwe and of the IPCC are not so different. Perhaps they are just different ways of knowing.

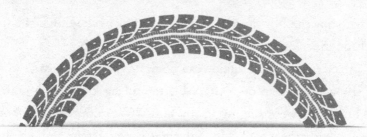

8

STEWARDS OF THE PRAIRIE

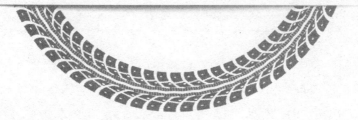

Back on the road, I opened my eyes in the morning to the tiny motel room and tried to remember where I was. Ah, yes: the Northern Lights truck stop in Spedden, Alberta. I began picking up the debris of yesterday's long day—it had been a couple of long days, in fact. The wet clothes had overwhelmed the ability of the air conditioner to dry them all, so many ended up hanging from the back of the bike during the day. No matter. The rolling clothesline would be ugly, but my limited wardrobe would dry quickly in the wind. Outside the door, across the way were the shoes of the Filipino family that runs the Northern Lights. I rolled my bike down the hall through a cloud of exotic spices.

Back on the road again, the boreal forest was drifting away. I'd made the leap onto the Alberta and Saskatchewan prairie. Munchkins were singing in my head: "We're out of the woods, we're out of the woods." Though there were trees around, much of the land was in wheat and soybean fields. I'd conceived the ride as a voyage across the ocean that was here hundreds of millions of years ago. Out on the grasslands, for the first time, I could imagine being at sea.

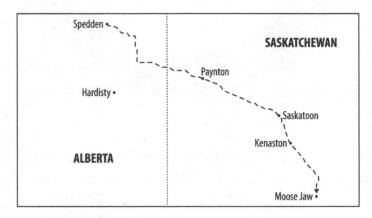

Spedden, Alberta–Moose Jaw, Saskatchewan,
riding out of the boreal forest.

With fewer trees came less shelter from the prairie wind. The days were increasingly defined by its direction: headwind, tailwind, crosswind. On the headwind days I tried to tuck down as low as I could, scratching out five, six, seven miles per hour. I would have taken the handlebar in my teeth if I could. Getting buffeted by crosswinds was a different kind of riding. I was sometimes tempted to lean into the wind, but then when the gust stopped I would have been falling out into the travel lanes.

Approaching a crossroads, I caught sight of a diner in the distance. Even though the deep forest was behind, it was still a long way between lodging, food, and civilization in general. I pulled in for an early lunch. As I was settling into my booth, a curious conversation was going on the next table over.

"Get there early you can pick your barrels."

I looked over to see an older man in full cowboy regalia. His embroidered shirt read "Canadian Finals Rodeo." I listened in as he was engaged in a favorite pastime of us old folks: complaining about young people.

"All these kids want to do is see how fast they are, never mind if they can handle a calf." Then on to another near-universal topic: people from out of town.

"They brought out some of these damn Syrian refugees. Don't speak English. Put 'em to work in the oil field. Then they get all grumpy and leave." In the emptying towns of the Great Plains, immigration would come up in conversation more than once.

I turned down Alberta Highway 41, the signs of which indicate that it's the Buffalo Trail. I didn't think much about it until, in the distance, brown dots covered a field on the horizon. As I approached, the shapes resolved into a buffalo herd. Perhaps a hundred grazed out in a field, safely on the other side of a fence. While cars didn't bother them, my sitting there on the bike clearly did. They gave me dirty looks (as much as I know how to interpret buffalo expressions), and then the whole herd decided to run away. Maybe not quite thunderous, but their hoofs certainly made a rumbling.

I came to recognize a certain rhythm out on the ocean. Days would unfold like an evening at the concert hall. The prelude was fog laying in the hollows, burning off in the early morning sun: flutes. As afternoon opens, puffy cumulus grew mountainous in the distance: drama, bassoons and bass violins. The third movement brought in the sweeping storms, lightning at first off in the distance, then nearby: climax, kettle drums and brass. Closing with the first evening stars, the sky washed clean: resolution, strings.

Settling into the motel rooms at night, the internet was a way to keep loneliness at bay, the link to family and friends. Losing Wi-Fi would have been devastating. On this part of the ride, evenings on the laptop were filled with transcontinental trash talk. I was working my way south and east to ultimately meet up with my friend and riding buddy Lynn Salvo in Moose Jaw, Saskatchewan. At sixty-seven, she's the Guinness Record holder for oldest woman to ride across the States, and this summer she was starting across Canada, west to east. She'd end up with that record too.

Lynn and I traded messages about our rides and who had the most excitement. Over in British Columbia, Lynn saw signs for avalanches and tsunamis. I countered with stories of black bears and oil trucks. She came back with a posting on grizzlies. Oh yeah? We got buffalo. It reminded me of a film from a few years back, *A Million Ways to Die in the West*. She talked about Category 1 ascents in the Rockies. Hey, I did a couple of Category 8s out on the prairie today. And in the Tour de France classification, that's *hors catégorie*. I had to admit, she had bigger hills and toothier predators.

I took a rest day in the prairie town of St. Paul to help my body and bike recover. The body's aches and pains would resolve on their own. The bike was another matter. My front derailleur wasn't working, so I had a very limited range of functioning gears I could use. And my rear tire was consistently losing air. With the nearest bike shop thirty miles down the road, I needed to come up with some repairs right there, in the motel. My noble conqueror of continents, my Horse of the TransAm, was a broken-down machine that still had to get me 800 miles down the road. I woke up at 3:00 A.M. worried about it. Then I fell asleep and dreamt about Paris. Funny how the mind compensates.

I woke up not in Paris, but with that bad boy still staring at me from across the motel room. I ate breakfast and attacked. The derailleur turned out to have a loose cable bolt. Taking off the tire, I pulled the tube out and re-inflated it. After filling the sink with water, I inched the tube through until I saw the subtle, telltale bubbles. I lifted the tube to my upper lip, where I could feel the tiny stream of air. There it was. Looking in the tire, I discovered sand and tiny pebbles. One pebble had gradually worn a minuscule hole in the tube. My last roadside repair had been in big wind, and blown debris had gotten into the tire. Bike work is easier in the morning light, away from the ominous worries of the small hours.

Heartened by the mechanical victory, I took a walk through St. Paul. On the eve of Canada Day (July 1), the town had banners celebrating the centennial of the end of World War I. One caught my eye: "Lieutenant Gordon Flowerdew, VC 1918," on horseback, next to a poppy. He joined the Royal Canadians in 1914 as a horseman from British Columbia. He was posthumously awarded the Victoria Cross for leading what would be the last cavalry charge of modern warfare. Canada did a good job of remembering the centennial, better than we in the States.

I settled in for dinner at a bright, cozy, farm-to-table cafe. The waitress had some time to talk, and I mentioned that I'd worked in the Louisiana oil rigs once upon a time.

"Guess I've had my time as an oil-field wife," she said. "His schedule was twenty-four days on and four days off. He'd come home and want the house all perfect and clean. Then he'd go out and raise hell. So that didn't last. But I love taking our boy up on the Icefields Parkway, in the Rockies by Jasper. I can see it all with new eyes, like for the first time."

I crossed into Saskatchewan on the Yellowhead Highway. This part of the province still has some oil development, as evidenced by the occasional pumpjacks out in the fields. But it's far from even. Outside Marshall, a tank reads "Black Gold Country!" and is topped by the skeleton of a Christmas tree. The town is virtually abandoned. The exclamation point is fading in the sun.

The highway runs parallel to the railroad, heading southeast. At the turn of the twentieth century, the trains were pushing across the prairie, with grain elevators and towns in their wake. I stopped by the village of Maymont, where the train line came through in 1905. That year May Montgomery asked her uncle, the railway construction company chief, to name the village Montgomery. He told her that he couldn't, because a town in Manitoba already had that name. So he took her first name and the first syllable of her last name to form the name Maymont.

The new prairie towns prospered. By the end of the war that claimed Lieutenant Flowerdew, grain was in high demand in a starving Europe cut off from the breadbasket of the Ukraine. A series of strong harvests in the 1920s added to the bounty. As the twentieth century wore on, mechanization and consolidation led to smaller demand for farm workers. The Great Depression hit the prairie hard in the 1930s. The towns began to decline. I rolled into one of those hard-luck towns and walked into a story of rebirth.

I'd had a rough night. I'd pulled into a bar/hotel in the farm town of Maidstone. An ancient pickup had been transformed into a planter full of peonies, with flags left over from Canada Day celebrations earlier in the week. I'd been dodging storms all day, and a big, black cloud was finally catching up. I had dinner in the corner of the loud bar where farmers and truck drivers were gathering over accelerating beers. The noise was oddly comforting. The second Molson was a mistake. I crawled up the stairs to the room and passed out by 7:30 P.M.

By the morning the rain gods didn't even make the pretense of waiting until the afternoon. Normally putting on my full rain gear makes it stop. I have cosmic powers that way. But my superpowers had declined, and I was wet starting at the first turn of the crank. I faced the prospect of fifty soaked miles before the first place to stop.

As happens on the road, the day didn't unfold as planned. About twenty miles out, I noticed a restaurant icon on the sign for the town of Paynton. I turned off the Yellowhead, and the pavement quickly turned to mud and gravel in town. Not a good omen. A few houses appeared, but no stores. I pulled up at a stop sign next to a pickup.

"Is there really a restaurant in town?" I asked the driver.

"Absolutely. Next left, then make a right on Main."

Perched on the quiet street, in several different shades of lavender, was the Purple Palace. The old square general store building appeared to be the only open establishment in town. It had a couple of castoff kid's bikes, painted in bright colors, leaned against the front. A sign reading DANCE LIKE NO ONE IS WATCHING was right next to one neon OPEN and two WELCOME signs. Well, if you insist.

I shook off and went in, peeling off my wet jacket and rain pants. The store had a few tables, but mostly was filled with antiques, teacups, exuberant decor, and wonderful smells coming from the kitchen. A lady appeared from the back in a blue-and-lavender top with purple streaks in her hair. I was detecting a certain leitmotif. She stepped out from behind the counter and gave me a big hug, pretty impressive with the amount of rain and road on me. She set me up with coffee, soup, and a sandwich. I was the only customer. Doris Frost, both waitress and owner, sat down to talk.

"This is an amazing place," I said. "How did it happen?"

She stretched her legs out. "I came here from Newfoundland, then Toronto. I was working as a care aid down the road in Maidstone until I retired a couple of years back. This store is 125 years old. Picked it up at auction for next to nothing. They tear everything down here in Saskatchewan. Nobody wants to keep anything."

"We have that problem in the States too."

"It was a mess, of course. It took me two years to refurbish it, with a lot of help from my family. It started out as an antique store, but then I wanted to make use of this big, old kitchen. So here we are."

I could smell the spices drifting off the lentil soup. "You're doing some nice work. And I take it you're fond of purple."

"You know, color brings out happiness. I want people to feel at home here."

"So who's your clientele?"

"The town's mostly retired farmers now. It used to be a lot bigger, but the trains don't stop at the grain elevators anymore. Everything used to travel by train. They have bigger elevators now, consolidated in just a couple of towns. But you know, we're starting to see some younger people in the shop, coming into town."

When I left, Doris had arranged for the rain to stop, and she even dialed up a tailwind. Flowers bloom out on the prairie, and sometimes in town too.

The next day the afternoon storms were building in the distance once again. I could sense cold air pouring into the tops of the thunderheads, cascading down through the dark interior, hitting the plains with a splash, and fanning out in all directions. The rush of air blew by me. The ragged edge of the line of squalls was not so far behind. It reminded me of an old Robert Johnson blues tune, "Hellhound on My Trail." I could speed up to try to outrun them, or slow down to let them pass by. But it was all an illusion of some kind of control. They had the run of the prairies, empresses, while I had just my narrow, little string of a road to balance on. The edge of the storm dealt me a glancing blow, some sprinkles and a little nearby crackling. She reached out to grab me, but I slipped away this time.

The next day brought heat, a headwind, and something new to the trip: a skyline. Above the soybean fields in the distance were the buildings of Saskatoon, the Paris of the Prairies. The Tourist Bureau of the City of Light might not be trembling in fear; Saskatoon is merely a delightful small city. Named for a local fruit, it's home to the University of Saskatchewan (the Huskies, since you asked). The large student population meant street fairs, quirky restaurants, and bike shops. I'd been missing all three.

I took a rest day to explore the town and keep all those promises I'd been making to the bike. One of the tires had a small chunk of rubber taken out. It still held air, but no sense pushing my luck. I visited the oldest bike shop in town, Doug's Spoke 'N Sport, for the luxury of having a pro work on the bike. Of course they had my tire size in stock. When I mentioned that I was on a thousand-mile tour, the mechanic moved my bike to the front of the line. He had huge forearms and gorilla hands, and installed the new tire without the benefit of tools. As his hands flashed over the bike, he weaved tales of winter rides on fat, specialized snow bikes with oversize tires. In a few minutes he popped the bike off the rack, shiny gears, tight brakes, and a clean new tire, ready for the second half of the ride.

The Hotel Senator was a grand, 110-year-old indulgence. Alas, the morning was quite loud: the hotel is next door to Winston's Pub, and England was beating Sweden in the World Cup.

About 40 miles out of Saskatoon, I stopped for a most remarkable lunch in Kenaston, once called the Blizzard Capital of Canada.

Now at least it's the Blizzard Capital of Saskatchewan. The Blizzards are their local minor league hockey team, and the town sports an eighteen-foot snowman in its center. The shimmering heat didn't quite seem to fit.

Once again, like the Purple Palace, I found a cafe that's a big part of what keeps a small town going. The Kenaston Cafe is run by Hai Dao, who came to the prairie by an amazing route. Leaving a refugee camp in Laos, he was sponsored by his uncle in the Yukon, where he went to high school. Going from the tropics to the Arctic suggests that Hai is made of strong stuff. He built RVs in Estevan, Saskatchewan, for many years, then came to Kenaston to take over the cafe. I was there with the after-church crowd in a little town where, as Hai said, everyone knows everyone.

And those blizzards? According to Hai, not what they used to be. Leaving the cafe, I saw a couple of kids who'd found a place out of the heat in the shadow of the Kenaston snowman.

As I rode across the plains, I thought a lot about what the cowboy said about the Syrian refugees, and I thought about the people I'd met: the outsiders, often immigrants, who seem to keep the place alive, the stewards of the prairie. In Avonlea, Saskatchewan, I'd meet a woman introducing the town to her Thai cooking in the local restaurant. The motels from Fort McMurray on south are run by Filipinos. When I'd get to South Dakota, I'd discover that one of the most technical oil field operations is being run by a group of Tongans.

The phenomenon of immigrants running things is certainly not limited to the northern Plains. In Storm Lake, Iowa, Art Cullen, the editor of the local newspaper, talked about his neighbors, the primarily Hispanic people who work in slaughterhouses and distill ethanol from corn. Many face deportation. He wrote:

> These Dreamers are our vitality, our future. They want to stay here with family, unlike so many of us who push our children off to Chicago or the Twin Cities. As our neighbors, they have prospered with our embrace. In our prairie pothole, a place glaciers left with natural abundance, nine out of ten students at the elementary school are immigrants—the schools here are a micro-city where you can hear thirty languages.[1]

Back home, I hear politicians say, "Our country is full." Out on the prairie, I didn't see it.

Ever since the start of the ride, I'd been pressing to meet Lynn Salvo, to roll into the same town on the same day. July 9 would be the day, with just seventy-two miles left to "Meet Me in Moose Jaw." I was on the road at sunrise, my long shadow stretched over the yellow canola fields. At the halfway point of the day's ride, I took a late breakfast break at the Twisted Sisters in Chamberlain before turning south. I asked one of the waitresses, perhaps one of the namesakes, what the next 35 miles had in store.

"Nothing," she said. "You've got plain nothing between here and Moose Jaw."

She was wrong about that. There was wind.

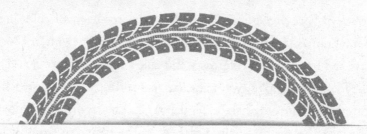

9

THE BOMB ON THE RIDGELINE

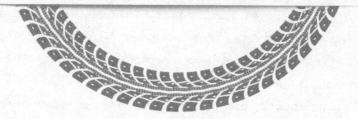

The oil truck behind me blasted its horn. I deserved it. I was off the bike and driving my rental car like an idiot, slowing down, looking for a place to turn off. My tires finally slid into the gravel of a little dirt road. I got my camera and walked out into a dry, stubbly grass field. There it was. Spread out before me was a vast tank farm perched on the ridge and into the valley below. I had trouble fitting the dozens of tanks in my viewfinder.

This was Hardisty, Alberta, a giant oil terminal and nexus where pipelines from the north, east, and west converge. Originally, Hardisty had been on my bike route. But back in the boreal forest south of Fort McMurray, I had lost a day to sheets of cold, horizontal rain. Of more urgency, I had a date to meet Lynn in Moose Jaw, so I had to bypass Hardisty and take a more direct route across the Plains. I found my way back the next spring. My jet approached Edmonton, the provincial capital and nearest airport, and I looked out over the gleaming city perched on the plains in front of the Canadian Rockies. I was returning in a petroleum-powered plane and connecting to a petroleum-powered rental car. Painted over the Edmonton baggage claim carousel: "This is Oil Country." Edmonton's hockey team is the Oilers. I was getting the idea.

Alberta is often compared to Texas. It's a big, brash, western province, with open, friendly people—witness Bryan, who had driven me around the surface mines in Fort McMurray. And there's a similar affinity for what drives the economy. If anything, Alberta's economy is more tied to oil than that of Texas. When asked at a Parliament meeting back in Ottawa whether other industries could one day take the place of oil, Rachel Notley, Alberta's premier, said, "Back home we ride horses, not unicorns."[1]

Before heading out to Hardisty, I spent a day wandering around Edmonton, trying to get a feel for the provincial capital. The sharply peaked Enbridge Tower, home of the namesake Canadian pipeline company, dominates the skyline. Having married a librarian, I have a certain fondness for her place of business, so I drifted into the main Edmonton library. Deep into the stacks, I discovered that I wasn't the first to take on a bicycle trip to this part of the world. In 2007, a group of nineteen young Canadian environmentalists rode from Waterton-Glacier Park on the US–Canadian border to Fort McMurray, documenting their ride in the book *Journey to the Tar Sands*.[2] Though the details of the rides were far different—I didn't camp and I didn't subsist on oatmeal and chili—the routes were quite familiar, as were the scenes of mines and tailings ponds. In many ways, not much has changed in eleven years. If anything, the footprint on the land has only grown. Yet there has been one brake on oil sands development since their ride. Pipeline capacity, and particularly the blocking of Keystone XL, has been a factor in limiting how much oil sands carbon gets into the atmosphere.

I drove southeast with the towers of Edmonton in my rearview mirror. I saw the forests of aspen gradually giving way to rolling plains. Not far outside of town, the first silver tanks of the Hardisty Terminal appeared on the horizon. This would be the storage point for the oil sands bitumen, to be transported south by the Keystone XL pipeline to refineries on the US Gulf Coast. The tanks I had driven off the road to photograph have a capacity of ten million barrels of oil. If the Keystone XL pipeline is to be built, Hardisty will be the northern terminus. It is also the terminus for Enbridge Line 3, the pipeline that would pass through the headwaters of the Mississippi and skirt Winona LaDuke's home on the White Earth Reservation in Minnesota. If Keystone XL is indeed, as James Hansen wrote, "the fuse to the biggest carbon bomb on the planet," this was what the bomb looked like on the ground, the phrase made real.

A little creeping paranoia was in the back of my mind. I had some reason to believe that I might not be all that welcome here. In the runup to the Alberta provincial elections, the local news was full of dark accusations of shadowy foreign environmentalists trying to bring the energy industry to its knees. In certain fading light, I might be considered a shadowy foreigner. Truth be told, I did feel a little like a spy.

Spy or not, Hardisty had been on my mind ever since the 2011 Keystone demonstrations in Washington. Keystone XL is the epicenter of one of the longest-running environmental debates in US history. Hardisty appeared on every map, in every article, in every TV spot on the Keystone controversy. I resolved to go there. Walker Evans, photographer of the Great Depression, once implored, "Stare. It is the way to educate your eye, and more. Stare, pry, listen, eavesdrop. Die knowing something. You are not here long." So in Hardisty, I stared.

As the phrase in Alberta goes, Keystone is to be a "route to tide-water" and the global energy markets. It would be the major route by which the carbon in the third-largest oil field on the planet gets out of the ground and into the air. Hansen's "carbon bomb" quote about Keystone drew howls of protest from the energy industry. Indeed, compared to global fossil fuel production, a relatively small amount of carbon comes from the oil sands.

But the fact of new pipeline capacity opens the door to huge increases in production. Two cases show this. When the controversial Dakota Access Pipeline was completed across the Standing Rock Reservation in North Dakota in 2017, oil generated from the Bakken shot up by one-third in less than two years. The pipeline was referred to as a "game-changer" by an executive for Continental Resources, a major driller in the Bakken.[3] On the other side, decreasing pipeline capacity slows production. Prices for heavy Canadian crude plummeted in 2018 after new production from the oil sands overwhelmed pipeline capacity, forcing Premier Notley to mandate production cuts.[4]

The tanks on the ridgeline are hardly a "bomb" in the literal sense. Tank farms are not uncommon in an economy that runs on fossil fuels. Then again, in one sense, the fossil fuel economy produced a very real bomb in 2013, in Lac-Mégantic, Quebec. One June night, on a hill above the town, a single exhausted engineer stopped his train, carrying seventy-seven tank cars of Bakken oil to the Irving Oil Refinery in New Brunswick. He set the hand brakes on the five lead cars and applied air brakes for the whole train. After he left, a small fire broke out in the lead locomotive. The fire crew followed procedure and shut down the locomotive. That also shut down the air brakes. In the small hours, after everyone had left, the train began rolling down the hill.

The train reached 65 miles per hour before it derailed and exploded in the town center. It destroyed forty buildings (half the downtown core) and killed forty-seven people. The blast radius was about one kilometer. After six years, the former downtown is still largely barren. The train line through town was the first thing rebuilt, and the oil trains still run. A rail bypass around the town won't be completed for four years.[5]

Lac-Mégantic is often cited to justify pipelines as a safer way to move oil than trains, yet pipelines have their problems. The 2010 Kalamazoo River pipeline breach mentioned earlier required five years for cleanup. A 2011 breach on the Yellowstone River at Laurel, Montana, dumped 63,000 gallons of oil into the river; eight years later, ExxonMobil paid a $1 million fine to "avoid protracted litigation."[6] These go along with the more recent spills on existing sections of Keystone, already mentioned.

Basically, this is seen as a cost of the business of moving oil. But towns don't get flattened and rivers don't get trashed by solar panels. The argument that we need to build an increasing number of pipelines to deliver an increasing amount of oil and gas goes against the view of the climate science community, which maintains that we need less carbon in the atmosphere, not more.

To step back a bit, the Keystone XL Pipeline has arguably been one of the biggest environmental issues in the United States over the last decade. How did it come to be so?

Proposed route of Keystone XL Pipeline,
with the solid line showing the existing Keystone system.

Keystone XL was first proposed in 2008. It was a big project, yet no one anticipated that opposition to it would be so fierce. Early resistance came from Native Americans and ranchers along the route—e.g., the Cowboys and Indians at the rally that I attended on the National Mall in Washington, DC. After the first environmental group filed suit in 2010, the Keystone proposal began to attract the attention of the larger environmental community. In 2011 some twelve hundred demonstrators were arrested in a civil disobedience action for sitting in front of the White House fence. A crowd of forty thousand gathered in Washington to protest Keystone in the winter of 2013, and a climate march brought four hundred thousand to New York City in 2014. President Obama denied a permit for the pipeline in 2015, while President Trump approved it in 2017, prompting a continued, extended legal battle.

The scientific argument for restricting fossil fuel development has formed the basis for much of this battle. With his knack for the controversial phrase, James Hansen again stepped into the discussion in a 2012 *New York Times* op-ed. He declared that if the oil sands and other "unconventional" sources of oil and gas, like the Bakken oil shale deposits, were fully developed, it would be "game over" for the climate.[7]

By that, he meant that the carbon contained in these unconventional sources would easily be enough to double the carbon dioxide concentration in the atmosphere from pre-industrial levels. His statement aligns with the more staid language of science documents. The 2018 report of the Intergovernmental Panel on Climate Change stated that global carbon dioxide emissions, of which fossil fuels are the primary source, need to fall by 45 percent by 2030 if the planet is to stay within a manageable level of warming.[8] Yet global fossil fuel emissions have *risen* by 6 percent since Hansen's statement.[9]

Like the runaway train in Lac-Mégantic, the fossil fuel industry is hard to stop, or even slow down. Who's behind Keystone shows why the fight has been so difficult. Charged with building it is TransCanada, a Calgary-based firm now renamed TC Energy. Who's financing it? Nearly all of the big global and Canadian banks, with JPMorgan Chase leading on the credit package to TC Energy. The organizations producing the oil certainly have an interest in getting more oil flowing through the pipelines. Although Suncor is the largest current producer in the oil sands, ExxonMobil is heavily invested, along with virtually all of the major international oil companies.[10] Basically, the biggest companies in finance and energy are invested in getting oil sands oil to market.

Not all of them, though. Up until 2019, the largest foreign leaseholder in the oil sands was Koch Industries, the second-largest privately held company in the United States and funder of a vastly influential political organization, Americans for Prosperity. That year, Koch sold its oil sands leases and backed out, following divestment from several other foreign oil companies.[11]

When Keystone was first proposed, support was bound to be formidable. So why did the environmental community choose to make a stand there? Wasn't it purely symbolic? Even in turning down the Keystone permit, President Obama remarked that "for years, the Keystone pipeline has occupied what I, frankly, consider an overinflated role in our political discourse."[12]

But the issue is more than just current production. Production of any individual oil project will be small compared to the global number. As the Dakota Access expansion demonstrates, bringing new oil supplies online will have a large impact on global carbon emissions. Starting any oil project means high capital costs. Once those costs are laid out and production starts, it costs far less to pump out the millionth barrel than it does the first. The effect is known as "carbon lock-in." Studies have demonstrated that new investments in oil exploration and infrastructure mean more carbon in the atmosphere for decades.[13] More colloquially, once the camel's nose is in the tent, it becomes difficult to keep the camel out.

In an era of accelerating climate change, the larger issue is the unquestioning acceptance of massive new fossil fuel projects. As climate writer David Roberts observed, criticizing the Keystone campaign for mere symbolism was "like criticizing the Montgomery bus boycott because it only affected a relative handful of black people. The point of civil rights campaigns was not to free black people from discriminatory systems one at a time. It was to change the culture."[14] For Keystone, the matter was one of denying fossil fuel companies their "social license" for unlimited production. That phrase would come up again in Alberta.

Approaching the motel in Hardisty, I noticed that I did stand out a bit. For starters, my rental car was the only non-pickup out front.

The woman at the front desk was outgoing, with curly brown hair.

"By the looks of your parking lot, I'm guessing that most of your business is with contractors," I said.

She looked up from her screen. "Oh yeah, we've got a ton of new work down at the Terminal."

"Can't place your accent. I'm guessing you're not from around here."

"Not even close. New Brunswick. Not much doing in timber and potatoes these days. Came out to where the jobs are." Drawn to a place to make a living, like people around the world.

After checking in, I took a stroll through the small town. Hardly anyone was on the street. A giant V of geese was working its way across a deep blue prairie sky. I walked past the hockey rink, ubiquitous in Canadian prairie towns. Much of Canada's identity is tied up with the cold, but that's changing fast. According to the Canadian government's most recent assessment, both past and future warming in Canada is, on average, about double the magnitude of global warming.[15]

At the edge of town, a little park looked out over the prairie. Fourteen flags were lined up, one for each of the thirteen provinces and territories to go along with Canada's maple leaf. A bench set on a little rise was next to an old horse-drawn plow.

Unlike Marshall, the town with "Black Gold Country" fading on the water tank, Hardisty isn't dying, but oddly for an active oil town, it's not exactly bustling either. A few vacant storefronts looked out over a wide Main Street with just a few cars. In the middle of the day, no one was on the street. I stopped in to the Hardisty diner. The waitress was talking with an older customer in a plaid shirt.

"We've got a great new homemade cheesecake," she said to him. He walked up, looked at the cheesecake, and made a throwing motion.

"For our ex-premier—Pow!"

From what I'd seen on television, I knew that this was provincial election day in Alberta. The polling station was open at the local elementary school. At stake was the re-election of Premier Rachel Notley versus her more conservative challenger, Jason Kenney. Not surprisingly in Alberta, the election was all about oil. More specifically, it turned on pipelines, not just Keystone but the Trans Mountain Pipeline, going from Alberta west through British Columbia. The government of British Columbia was staunchly opposed to the latter project, though a fair amount of the opposition came from the cities of the east, particularly Toronto and Montreal.

What had been a surprise four years earlier was Notley's election, from the relatively less conservative New Democratic Party (NDP). In a partial exchange for support for Alberta's pipelines, Notley had agreed to a provincial carbon tax, i.e., putting a price on greenhouse gas emissions, as advocated by the national government of Justin Trudeau. Trudeau's government later purchased the Trans Mountain Pipeline to make sure that it would go through. (At this writing Trans Mountain, like Keystone, remains held up by the courts.) Notley also supported doubling subsidies for petrochemical projects.

By contrast, Kenney attacked opponents of the oil sands robustly. He advocated "turning off the taps" of oil to British Columbia for their pipeline opposition. He would remove the cap on oil sands greenhouse gas emissions and halt the shutdown of Alberta's remaining coal-fired power plants. Kenney said he would launch a public inquiry into what he believes is the large-scale foreign funding of anti–oil sands campaigns. In announcing the inquiry, he said, "For more than a decade, Alberta has been the target of a well-funded propaganda campaign to defame our energy industry and to landlock our resources."[16] That was a good reason for me to keep a low profile around Hardisty.

One of the biggest newspapers in the province, the *Edmonton Journal*, stated in its endorsement of Jason Kenney that oil would "bankroll a greener future":

> NDP Leader Rachel Notley's carrot approach on the pipeline file was a well-meaning endeavor, but it failed to produce results. Now, it's time to try the stick. . . . [The NDP] made gutsy moves, like bringing in a provincial carbon tax, an earnest effort to fight climate change. But that approach was supposed to be the ticket to winning social license for Alberta's oil. It should have been our route to tidewater [international export markets].[17]

That "social license" phrase popped up again. In fact, it seems to be what the whole fossil fuel debate is about. New big oil and gas projects require government approval, which puts them in the public sphere. People have to believe that the jobs and wealth that fossil fuel production generates are worth the environmental destruction. If that social license is lost, pipelines aren't built, drilling doesn't happen, and the massive investment machine that keeps it all going freezes up. The reserves in the ground become stranded assets and the valuation of oil and gas companies plummets.

Nothing like that was going to happen on Alberta's election day, April 16, 2019. Based on local television reports, Notley and the NDP were in trouble. The day before in Calgary, 1,100 office workers had an "empty the towers" demonstration, leaving their office buildings in a pre-election push for oil and gas. When the final results were in, Kenney and his United Conservative Party had won a solid victory.

As promised, shortly after they were sworn in, Kenney and his colleagues repealed Alberta's carbon tax. In May 2019, they were set to celebrate the repeal. But ironically, three years after the Fort McMurray fire, Edmonton itself was blanketed in wildfire smoke. The festivities were canceled so that Kenney could attend an emergency fire briefing.[18]

It wasn't clear how much Alberta policy regarding pipelines would change in practice following the election. By 2017, well before the government purchased it, the Alberta deputy premier was already referring to the Trans Mountain as "our pipeline."[19] The influence of the oil industry goes across government and academic sectors, as described by Kevin Taft, a former member of the Alberta Legislative Assembly:

> In Canada, top civil servants "fully aligned" their departments' priorities with the oil industry; government scientists were muzzled; and leaders of university institutes (some on the payroll of industry) redirected institutional focus to the priorities of the oil industry.[20]

That influence certainly extends to the United States. In the summer of 2019, the Smithsonian Museum of Natural History in Washington, DC, opened its refurbished fossil exhibit, entitled

"Deep Time." The exhibit was greeted by big crowds and strong reviews. In describing climate in the "Age of Humans" section, the narrative does describe that "we're changing the planet faster than any other species in Earth's history." But it's curiously quiet on the relative role of fossil fuels in driving that change. That may have something to do with its namesake: It is the David H. Koch Hall of Fossils.

On my last night in Hardisty, I got a quick lesson in east–west contrasts in Canada. Much of the commentary against pipelines and "dirty oil" comes from Quebec and Ontario. As the local paper put it, "Albertans are famously averse to people out east telling them what to do."[21] I walked across the highway from my motel to The Leaf Sports Bar. It was Paralyzer Tuesday, a reference to one of their specialty drinks. Red Bull was also on the menu, with the tagline, "You were here too late last night, and we love you for that." Paralyzer Tuesday also was hockey night, and the Stanley Cup playoffs to boot. As Canadian as it gets. The crowd in the back of the bar was undoubtedly dominated by people working at the Terminal. I settled into a side table near the front.

At the bar, an older man was nursing his beer. His T-shirt showed a skeleton riding a Harley trailing flames. In Gothic lettering below: SONS OF ARTHRITIS. To the left, a younger man was pleading with the woman behind the bar, apparently his boss at the Terminal in their day jobs.

"Promise, I'll be at the front gate at 6:30."

"More like 6:34," she sniffed.

The hockey playoff game was close and entertaining. The Toronto Maple Leafs had scored the go-ahead goal on the Boston Bruins. I had assumed that the crowd would be behind the Leafs—after all, it was the name of the bar. All was right with the world. At the horn, Toronto held on for a 3–2 win. As the Leafs lined up to congratulate their goalie, a loud voice came from the back of the bar: "Fucking socialists!" Back in the States, that epithet might be aimed at Boston. But that wasn't who he was talking about. I paid the tab and slipped back across the highway.

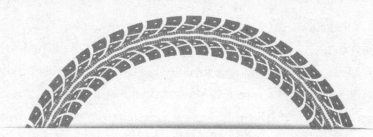

10

THE BIRDS OF SASKATCHEWAN

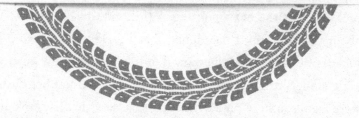

T he dirt road stretched out to a treeless horizon below an impossibly big sky. Shadows of clouds painted dark patches on the tawny landscape. Stubble of old winter wheat and canola covered the southern Saskatchewan fields. Mid-April was bringing the first hints of spring to the prairie. Small white flowers were emerging intermittently. The pickup seemed to float along the gravel at speed, with only a tenuous connection to Mother Earth. In the distance, the linear skyline developed ripples: the Big Muddy Badlands.

I had reason to be nervous. It wasn't my truck, it was Al's. Al sat next to me, occasionally pointing out some aspect of the prairie or its birds, subtleties invisible to a city kid like me. At least I didn't have to worry about traffic. We saw three other cars that afternoon. Al says that in this part of Saskatchewan you wave when one passes, just to be friendly. We were some distance from the oil fields to the north and south, but like most places on the planet, not far from their influence.

I had met Al Smith purely by serendipity ten months before. I had arrived by bike in Williston, North Dakota, on a hot July day (see Chapter 14). I was settling in to the sumptuous Motel 6 breakfast bar the next morning. While talking to another man about the bike trip, I mentioned that I'd ridden down to North Dakota through Saskatchewan. Randi, a gregarious woman with a big smile sitting next to us, piped in that she and her husband were from southern Saskatchewan. That was my first look at Al, a wiry man with a cane. He had a goatee and, behind glasses, a twinkle in his eye. His hair tended to go off in multiple directions simultaneously. Al, a retired ornithologist, is one of the leading authorities on Saskatchewan birds and their migration patterns. He began telling me a tale of birds and oil:

> I spent most of my career with the Canadian Wildlife Service. Back in 1971, I almost got fired over the tar sands. *[Yes, he calls them the tar sands, and often gets corrected in Canada.]* I spoke to Grant Notley, who was a member of the Alberta Legislative Assembly, about waterfowl up in Fort McMurray landing in the tailings

ponds and never getting out. You might remember that his daughter Rachel ended up as Alberta Premier. Anyway, he went on TV to talk about the birds getting caught in the ponds, then blabbed my name as the source. That was a firestorm, and I nearly lost my job. Despite the fact that thousands of birds *were* dying. But my indiscretion may have helped push the regulations. It could be one reason that the companies started putting up sound cannons and raptor decoys around the ponds to keep the birds out.

Al and Randi and I talked and, as the morning passed, we scratched email addresses on the backs of cards. As a result of shared emails over the winter, an invitation to visit their home in Saskatchewan dovetailed with the trip to Hardisty. Leaving the tank farm behind, another flight brought me to Regina, Saskatchewan's capital. Beyond Regina, the land is pure and flat, more like that ancient ocean.

After an hour's drive, white grain bins in the distance announced the presence of Avonlea, where Al and Randi live. Avonlea is a bit over a hundred years old, named for the fictional town in the children's classic *Anne of Green Gables*. Al's directions should have prepared me for the landscape in a town of four hundred.

"We have no address," he wrote. "Just the gray house opposite the rink." I had landed on the prairie electronically the month before. My Google Street View cyberscout found the ice rink. Sure enough, there was a gray house across the street.

Avonlea reminded me of another place closer to my home. There is a low isle in Chesapeake Bay by the name of Smith Island, inhabited by a community of watermen since colonial days. In the present day, it's eroding, steadily losing ground to sea-level rise. To the north of Smith lie Marsh and Bloodsworth Islands, places where towns used to be, now salt marsh, silent warnings of what is coming. Avonlea's place on the plains is like that, but for a different reason. Up and down the Great Plains, the towns have been emptying over the years, victims of bigger farms and more mechanization. To the south of Avonlea is Kayville, now nearly abandoned. Al would later point out the old watering hole Dodge City.

"Great bar. Shame we lost it," he said as we drove through.

Randi and Al welcomed me to Avonlea with dinner at the local restaurant and tavern. It was nothing fancy: linoleum tables, hockey memorabilia on the walls. As Randi said, "I love dive bars." The surprise was a delicious Asian menu, courtesy of the owner's Thai wife. In Avonlea, as in Fort McMurray and across much of the Canadian prairie, the Southeast Asian diaspora plays no small part in keeping the towns going.

My mornings with Al were lessons. A colleague had written about Al that his manner was to "teach others about the birds and find ways to defend them and their habitat." It was a desperate, short, crash course, interspersed with commentary on the Montreal Canadiens, the Habs, his team. Al never goes out without his Habs hat, and fortunately for me, they were out of the Stanley Cup playoffs by the time of my visit. I asked Al about how he first got into ornithology:

I started out as a kid from the wrong side of the tracks in Saskatoon. I got into university, but I wasn't a very good student. Afterwards, I got a job with the Wildlife Service, and it seemed that there were a lot of people that were just putting in time. But in working with birds, I found something that I enjoyed doing, that was important, and where I could make a living. What more could you want? I was never happier than out in the field in the Arctic with twice-a-day radio contact, surveying birds and talking with other members of the crew.

The morning's lesson began with a bird that Concetta, an avid birder, had seen near our home in Maryland: the blackpoll warbler. Weighing about the same as a ballpoint pen, the tiny black-and-white bird feasts in summer on the plentiful bugs that inhabit Canada's boreal forests, some of the same bugs that feasted on me when I got off the plane in Fort McMurray. Fall comes early in the north, and the blackpolls must find their way south and east across the barrier that is the Great Plains of Saskatchewan. Their weight provides a clue to how they make it across. Al runs the Last Mountain Bird Observatory north of Saskatoon, where they catch, weigh, band, and release many of the migrating birds coming through. The blackpolls put on little bursts of fat every day, gaining enough to fly the following night. The data suggest that they are making hopping flights across the plains, pausing to feed using whatever food source is available.

"But sometimes what birds do will completely surprise you," Al said. As they approach the East Coast, the blackpolls slow and put on a remarkable amount of weight. They linger. Sometime around the autumnal equinox in late September, the Bermuda

High, the great ridge of air that delivers the summer's hot, sultry southwest winds, breaks down. The sky opens up to the big blows of the fall, the cold north winds. In my water, the Chesapeake Bay of Maryland and Virginia, these strong winds mix oxygen into the depths, opening the deepest reaches to our famous animal, the blue crab. Up in the air, the blackpolls get their signal. With the north wind behind them, they make their break for another continent.

The Information Age has brought us trackers tiny enough to attach to the leg of a warbler. What they reveal is wondrous. The blackpolls leave the East Coast somewhere north of North Carolina to fly roughly eighty hours nonstop over the open Atlantic, some 1,600 miles, ultimately reaching their winter grounds in South America.

I wasn't quite sure where we were going that afternoon.

"Are we going to be on any dirt roads?" I asked plaintively. "Not sure if I want to take my rental car out if we are."

Al looked down over his glasses. "Pretty much all of it. Don't worry. We'll take my truck."

So we piled in and rode off the island that is Avonlea. As the grain bins faded off in the distance, the prairies spoke softly, not in the bold tones of a redwood forest or a sheer rock face. Al looked out to the horizon.

"I love that particular shade of brown that is native prairie. I remember once sleeping out there. What woke me up was a couple of vesper sparrows fighting on my sleeping bag. Guess I was part of their territory."

I drove him around the dirt roads through the Badlands. A recent stroke had hurt him, and for now he couldn't drive. His muscles were stiff and painful, and he was constantly stretching.

"The brain's rewiring things, but the body's resisting, as though it can keep things the way they were. The doc says that battle is what's causing all the pain. I have good days and bad days."

"Is this a bad day?" I asked.

"Any day I can get back out in the field is a good day."

Not too far out on the plains is a tree, probably part of a shelter belt, where trees were planted as windbreaks on a long-abandoned farm. From the truck, Al pointed out the great horned owl's nest.

"You can get a closer picture if you want."

At first, she was hard to pick out, perfectly camouflaged in the dry branches. Then it was as though she jumped out at me. I recognized a huge face with two slit eyes, motionless and ominous. I tried to tiptoe up discretely, despite being the sole vertical figure on the prairie for dozens of miles. I had a funny feeling that she saw me. I got my photograph, with the fake shutter sound helpfully unmuted. Back at the car, Al mentioned, "They normally won't attack until dusk or dawn."

Wait, what?

"Yeah," Al added, "when people are checking on nests with young, they usually wear helmets and long leather gloves."

"Um, so you think she has young now?"

"Probably."

Al loves maps. His topography sheets of southern Saskatchewan are pawed over, frayed at the edges, and speckled with highlighted nest locations. The ferruginous hawk, a grassland raptor, is his favorite bird, a high honor. He's documented over two thousand nestings of the hawk across Alberta and Saskatchewan. I asked what was so special about this bird.

"The ferruginous hawk is the largest and most handsome of the buzzard hawks, the buteos. Unlike the red-tailed and Swainson's hawks, which have harsh calls, the ferruginous has a soft call that gets carried off by the wind, as if evoking the spirits of a bygone era when there was nothing but gorgeous prairie. I never get tired of seeing one."

Al was nervous this day. Migration patterns are changing, and I think he was afraid he might not see his old friends. We had been to three former ferruginous nests without seeing one.

"They should be here by now," he said. Then, coming out of a draw, on a hillside: the noblest buteo of them all. The huge, brown-winged, white-breasted hawk scanned the prairie from the grass of a fenced-off field.

"I can't climb over the barbed wire, but you can give it a try," he said. Why did I feel like I was getting set up again? As I delicately made my way over the top strand, I heard the sound of wings and felt the rush of air as the great bird lifted into the sky.

We headed south, toward the border with the States. Before too long, the indentations on the horizon grew bigger, and we entered the Big Muddy Badlands. They aren't quite as spectacular as the badlands of the Dakotas, but they're even more remote. We stopped for lunch at the base of an isolated pinnacle called Castle Butte. Our legs hung out of the back of the truck, the only place for miles out of the wind. Dust devils spun on the road. Kestrels—small falcons—turned and played in the torrents of air.

Tipis on the National Mall, April 2014.

Dump trucks hauling oil sands, Syncrude plant, Fort McMurray, Alberta.

Tailings pond, Syncrude oil sands project north of Fort McMurray, Alberta.

Shoreline of a tailings pond.

Sulfur mounds, Syncrude plant.

ABOVE: Pyrocumulonimbus cloud, May 4, 2016, looking north, 30 km from Fort McMurray. *Photo courtesy of Alberta Agriculture and Forestry.* BELOW: Wildfire near Anzac, Alberta, May 4, 2016. *Photograph by Chris Schwarz/Government of Alberta via Creative Commons.*

Oil tank outside Marshall, Saskatchewan.

ABOVE: Tish Boychuk and Doris Frost of the Purple Palace. BELOW: Teenagers escaping the heat in the shadow of the Kenaston snowman, Kenaston, Saskatchewan.

Tanks at Hardisty, Alberta, proposed northern terminus for the Keystone XL Pipeline.

ABOVE: Is this what an angry owl looks like? On the prairie, southern Saskatch-ewan. BELOW: A red knot (center, with orange breast) amid thousands of shore-birds, Mispillion Harbor, Delaware. *Photo courtesy of Concetta Goodrich.*

"The climate system is an angry beast, and we are poking it with sticks."
—Wallace Broecker. Drilling rig outside Williston, North Dakota.

ABOVE: The Old School, Fortuna, North Dakota.
BELOW: Prairie Lighthouse, Fortuna, North Dakota.

ABOVE: Gas flares off oil wells, northern approach to Williston.
BELOW: On the horizon, gas flares off wells on a dusty day outside
Watford City, North Dakota.

Concetta among the Teddys, downtown Medora, North Dakota. From left, actors in costume portraying Quentin Roosevelt (Austin Artz), Teddy from Yosemite (Larry Marple), President Teddy (Gib Young), and Teddy's aide Major Archie Butt (Matthew Burr).

Roosevelt in buckskins, 1885.
From the Library of Congress.

Puck cartoon from 1904 depicting George Cortelyou, Roosevelt's Secretary of Labor and Commerce, "putting the screws" to the Trusts. *From the Library of Congress.*

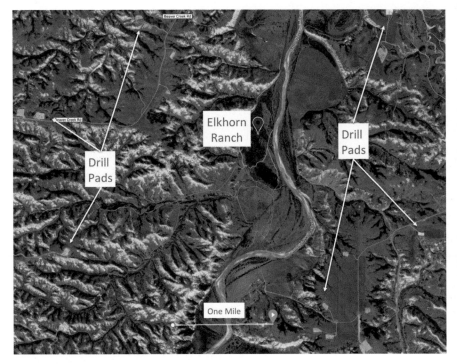

Site of Roosevelt's Elkhorn Ranch, North Dakota, and nearby oil drilling pads.
Satellite view from Google Maps.

"You know, they look like hawks, but they're actually parrots that have evolved to hunt," he said. "They came to live in the same places as hawks, and they ended up with many of the same characteristics."

I dug deep into my memories of college biology and stretched.

"Convergent evolution?" I asked. Al nodded with the satisfied look of a teacher.

We finished our lunch watching the kestrels.

"I always enjoyed the science, trying to use different tools to learn where these birds come from and how they live," he said. "But it's a field that has a certain amount of art. It's a science where you can hold beauty in your hand."

Back at the house that night, Al left a book on my bed called *The Narrow Edge* by Deborah Cramer. I asked him about it.

"That's about one of your champion migrators. This one's pretty amazing, and your neighborhood has a big part in the story."

The red knot, a robin-sized shorebird, spends the northern winters on the tidal flats of Tierra del Fuego, at the southern tip of South America. It then flies 9,500 miles, much of it over open ocean, to journey to the other end of the planet. It reaches its breeding grounds in the Canadian Arctic in May and June, often passing by Al's territory in Saskatchewan.

The key stop along the way is Delaware Bay, not far from my home in Maryland. The red knots arrive at the Bay in mid-May, famished from their journey, just in time for the arrival of the horseshoe crabs. The ponderous, dinner-plate-sized crabs are living fossils, surviving since before the time of the dinosaurs. Delaware Bay has the highest concentration of horseshoe crabs of anywhere in the world. During the May high tides, they lumber up onto the beach to mate and lay green masses of eggs in the sand. The knots and other shorebirds are waiting. Birders refer to May on the Delaware as the Feast on the Beach. The epic travelers spend a week or two fattening up on crab eggs for the rest of their journey, nearly doubling their weight.

"They show up at the end of May here in Saskatchewan," Al told me. "Your part of the world and mine are connected in more ways than one."

The red knot, like many birds, is endangered by climate change. In fact, it's the first US bird listed as threatened under the Endangered Species Act explicitly because of changing climate.[1] The threats come in a number of ways. Warming in the Arctic is reducing its nesting area. Ocean acidification, caused by dissolution of carbon dioxide in water, is reducing the ability of shellfish to create their shells, affecting one of the primary food sources on the bird's long journey. A rise in sea level reduces the area of the tidal flats, upon which the red knots rely to feed. What's happening locally, in addition to global climate change, matters as well. Both Delaware and New Jersey have instituted either bans or restrictions on horseshoe crab catches, hoping to keep this key food source from disappearing. Whatever the cause, since 2000, the red knot's population at key stopovers has declined by roughly 75 percent.[2]

It's often said that birds are part of an early-warning system, sending us messages about climate change and the degrading quality of our agricultural and industrialized landscapes.[3] The warnings may not be so early anymore. For birds, one simple idea is that they're well-suited to a changing climate, as they can simply move their range. But many species are closely tied to a certain food source, and that food may not be able to change its range. The mismatch could be deadly.

We can consider how ecosystems may respond to a new climate. It could be that the most adaptable birds—like starlings and house sparrows—will flourish in this new world, while the birds with intricate, interdependent life cycles will disappear. In the oceans, because of acidification, some speculate that ecosystems will come to be dominated by creatures without exoskeletons or shells, like jellyfish. In the sky, perhaps we are seeing the permanent decline of intercontinental voyagers like the red knot and the blackpoll warbler.

The decline of the red knot is against the background of a larger loss of birds. Recent research documents that since 1970, over a single human lifetime, roughly one in four birds has disappeared from North America. Just since 2007, based on US weather radar data, the amount of "bird biomass" flying over our heads has declined by about 14 percent.[4]

As the United Nations, and Al, would say, while climate change exacerbates the loss of birds, habitat loss has been a huge factor. Al talked with me about how central Saskatchewan came to look the way it does.

"The shelter belts, the natural sloughs, and the prairie potholes, these were places where birds like the blackpolls could rest and feed and refuel. But you don't need them for the giant farms, so they just mindlessly cleared them all. From Moose Jaw to Regina, there's not a lick of natural prairie left. If you want to make it look like a desert, it will act like a desert."

Al shook his head.

"Had a colleague from Venezuela come to visit. Folks from the north rag on them for not preserving the rain forest. Looking out on our plains, I was thinking about what he didn't say, what he was too polite to say: 'Couldn't you have saved more?'"

The night before I left, Al and Randi invited several of their farmer friends over for dinner. Al had mentioned that I was visiting, and that I'd worked in climate science. I asked about the long-term health of farming in Saskatchewan. One farmer, owner of one of the largest farms in the area, spoke up.

"Twenty years ago they said we couldn't grow enough to feed the world. Now we're making too much."

The after-dinner conversation turned to politics, and particularly the Alberta provincial elections to the west, which had just taken place. There, the electoral debate had revolved around the recently instituted carbon tax and the role of Prime Minister Trudeau. The PM wasn't popular around the table. One farmer raised a finger.

"Name one good thing that Justin Trudeau has done." He smiled a little as he turned to me. "I suppose that you would say the carbon tax."

"Well, yes," I said. "But if you don't like the carbon tax and you believe climate change is a problem, what would you suggest we do about it?"

"I think that things may be going to pieces regardless. I'm not sure there's much we can do. It's kind of like a ship that can't be turned."

I winced, of course. That would be one of those self-fulfilling prophesies. From my perspective as a climate scientist, I can pretty well guarantee that if we do nothing—burn fossil fuels at the rate we've been going—it's not going to be pretty. There's no reason to believe that the heat and the fires and the sea level rise will slow down on their own. Al talked with me after everyone had left.

"Bear in mind that these farmers use a whole lot of fossil fuels. I talk with them often, and we get into it from time to time. They do a lot of damage to habitat, and they use a ton of chemicals. But we need to listen to them. They're trying to make a living in a difficult line of work. Not to mention putting food on the table."

In some ways, Al and I are in analogous positions. I feel about people working in the oil patch the way Al feels about farmers. These are people we need to convince about the need for change, while recognizing that people need to make a living.

As in many things, younger people have a different perspective. It's not quite as easy for them to shrug and say that things can't be changed. Some weeks after the Avonlea visit, an Ontario fifteen-year-old's letter to her member of Parliament was cited in support of a climate emergency resolution in the House of Commons. Elizabeth Rose said, "People don't really understand that for the kids, it will affect them so much when we grow up and this will just be our lives."[5]

As we were cleaning up from dinner, I noticed a piece of crystal and a large atlas-sized book on a table in the corner of Al and Randi's dining room. The crystal, on closer look, was the award for regional ornithology from the American Birding Association. Al cradled it in his hands.

"Same award went to Roger Tory Peterson."

"Of the Peterson bird guides?"

"Yep. Guess I did okay."

I reached to open up the atlas. It was huge, beautiful, and intricately illustrated, the product of a small army of volunteers. The lead editor was Alan R. Smith. It was *Birds of Saskatchewan*.

In a postscript to the Avonlea trip about a month later, Al's words led to another place, far from Saskatchewan. Concetta and I found ourselves squinting through binoculars and long camera lenses, looking out across Mispillion Harbor, on the shores of Delaware Bay, five days after a full moon. Shorebirds flitted across the sky, but thousands more gathered on the island beach across from us. At the water's edge, the line between sand and water seemed lumpy. Zooming in carefully revealed the round silhouettes of hundreds of horseshoe crabs. The Feast on the Beach was in full swing.

We fired off dozens of photos—for birders, it truly is shoot first and ask questions later. Concetta, who has an eye for birds, could identify dunlins and ruddy turnstones and various sandpipers. But not the distinctive shape and coloring of the red knot. Then, slightly larger than its neighbors, an almost theatrical appearance: a knot stood as a face in the crowd. After a four-day, nonstop flight from South America, a little party was in order.

Later, scanning though our photos, we would find red knots here and there, but never in abundance. We went to one beach after another on both the Delaware and New Jersey sides of the bay, only occasionally catching glimpses. On some beaches, feeding grounds were cordoned off with yellow tape, keeping tourists like us at a distance, looking to protect their long-distance guests. There's good reason to be worried. As recently as 2000, some fifty thousand red knots were counted in Tierra del Fuego. In January 2018, the number had dropped to 9,840, the lowest since surveys began.[6]

Not long before our visit to Delaware Bay came a report from the UN Intergovernmental Science-Policy Platform on Biodiversity and Ecosystem Services.[7] This assessment, authored by scientists from around the world, warned that around one million species of animals and plants are threatened with extinction, many within decades. The vast, ominous nature of the report somehow made it easier to ignore. Perhaps it has something to do with a quote attributed to Stalin: "A single death is a tragedy; a million deaths is a statistic." So I prefer to represent the issue with one small, epic traveler who journeys literally to the ends of the earth, and the hope that we can bring the planet around to a place where such creatures might be able to survive.

After all, as Emily Dickinson wrote, hope is the thing with feathers.

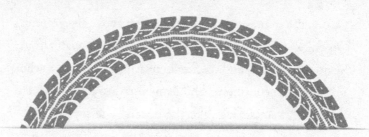

11

WHERE THE SEA USED TO BE

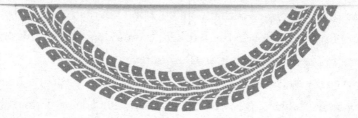

When I returned from Canada, I got back to research and following stories down rabbit holes. The bike was in the shop for some long-postponed love. A late season nor'easter was spinning up off the coast, and I was buried in a stack of books and magazines. A *National Geographic* article about the North Dakota oil fields caught my eye. Susan Connell, a truck driver in the Bakken, stood on a well site at night. Her job was to haul off the salty wastes flowing back up from fracked wells:

Connell's standing on a catwalk, high above the ground, opening the hatch on a tank of clear salt water that came from thousands of feet beneath the surface, in the middle of a continent. She leans forward and breathes deeply. "It smells just like the ocean," she says.[1]

The smells rising from oil field brines are usually anything but bracing. Typically, the mixtures contain a cocktail of drilling chemicals and the rotten-egg smell of hydrogen sulfide. Being overcome by fumes is one of the many hazards of working in the Bakken. But her comment made me think about the great inland sea that had formed both my starting point in the Alberta oil sands and my destination in the Bakken. Susan Connell was inhaling a sea breeze from four hundred million years ago.

On my shelf sits a favorite Rick Bass novel about geologists hunting for oil in Montana, *Where the Sea Used to Be*. It made me curious about this ancient sea, the two giant oil fields that it birthed, and the vast plains in between. Putting together the geologic jigsaw puzzle of how this oil came to be involves the notion of eras strikingly different than our own, together with some violent swings in climate. Along the track of the bicycle journey, only a few thousand years ago, lay the largest lake on the planet. On the way, some very human, eccentric, and brilliant characters helped assemble the puzzle pieces.

In looking at the earth's history, numbers over a few thousand years don't work very well in the human brain. In considering eras so long ago and so different from our own, ages of tens to hundreds of millions of years tend, in John McPhee's words, "to awe the imagination to the point of paralysis."

For the purpose of this journey, let's consider just three periods as relevant. The first, the Devonian, about four hundred million years ago, is known as the Age of Fishes, when vertebrates first began to emerge from the ocean, long before the dinosaurs. This was the ocean that Susan Connell was smelling. It's the age that gave rise to the Bakken shale. My original, simple idea was that, in a ride through the boreal forest and the Great Plains, I would be traveling across the basin of the same ocean that created both the Bakken and the Alberta oil sands, perhaps two shorelines on the same inland sea.

In fact, the oil sands were formed far later: during our second period, the Cretaceous, 65–145 million years ago. Fossils provide evidence for what the Cretaceous was like. Duck-billed hadrosaurs wandered the landscape, and T. rex may have been as abundant as hyenas. The climate was steamy and swampy. In fact, my teenage brain distinctly remembers the Cretaceous, with Raquel Welch battling dinosaurs in a low-cut cavewoman outfit in the 1967 movie *One Million Years B.C.* I believe I was focused on Ms. Welch, but it was definitely hot. More to the point, as the most recent prolonged warm interval in earth history, the climate of the Cretaceous can provide an analog, a place where the planet may be heading once again.

Our third period is far more recent, and colder. The Quaternary extends from two million years ago to the present. By this time, the shallow saltwater sea was long gone from the center of North America, and a huge volume of the planet's water was locked up in ice sheets. As these glaciers were melting and receding, a giant freshwater lake formed that would influence the global climate in a remarkable way. More on this later.

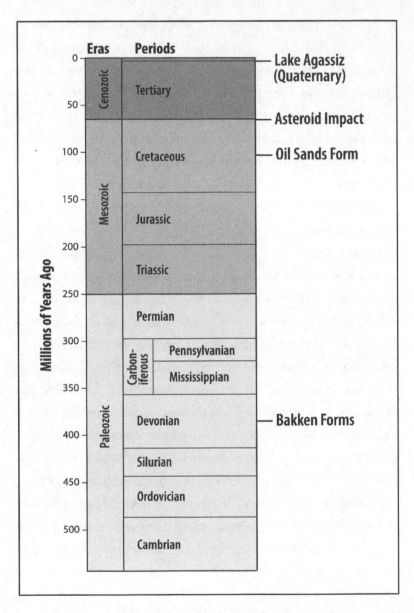

*Geologic timeline and origin points for the Bakken formation,
the Alberta oil sands, and the glacial Lake Agassiz.*

Like all oil fields, the two that bookended my ride—the Alberta oil sands and the Bakken—were formed from carbon taken out of the ancient atmosphere. In the ancient sea, microscopic plants known as phytoplankton used the sun's energy to form hydrocarbons from carbon dioxide dissolved out of the atmosphere. The inland saltwater sea that covered today's Great Plains was known as the Western Interior Seaway during the Cretaceous, the period when the oil sands formed. It was fertile, with phytoplankton packing away this principal greenhouse gas. When the plankton died, they fell to the bottom, creating carbon-rich sediment layers. Over the eons, they were buried under other layers, and the weight of those layers created heat and pressure. After more than a hundred million years in this pressure cooker, the remains of these organisms were transformed into oil and gas, hence the term fossil fuels. The energy for the gas flames that we burn today comes from the light of an ancient sun.

These hydrocarbons tend to move through pores in the rock, and they're often found far from where the phytoplankton deposits were first laid down. They migrate from high pressure deep underground to lower pressure near the surface. Sometimes oil bubbles up to the surface, as in the oil seeps off Santa Barbara, the oil springs in Azerbaijan, or the opening credits of *The Beverly Hillbillies*. More typically though, oil gets trapped below impermeable geologic structures. It can then form underground reservoirs, some the size of cities. Finding and exploiting those reservoirs is what the oil and gas industry is all about.

The older of the two great northern oil fields, the Bakken, originated in this Age of Fishes, the Devonian. It's a saucer-shaped basin two to three miles down centered around Williston, North Dakota. The Bakken oil is trapped in shale and known in the business as "tight oil." Tight oil is closely bound to the rock and only accessible for extraction if the shale is pulverized through the process of hydraulic fracturing.

The Alberta oil sands were formed in a similar way during a much later period, the Cretaceous. Throughout the Cretaceous, hydrocarbon-bearing sediments had been laid down, buried, and subjected to the intense heat and pressure that generates oil and gas. At about 110 million years ago, the sands of the McMurray Formation, what would become the oil sands, were laid down by the rivers that once drained into the ancient sea. At a Fort McMurray mine site, the giant shovels have uncovered fossilized wood near the bottom of the oil sands layer. The trees were an ancient species of conifer abundant during the Cretaceous, when the original sands were deposited.

Oil migrated along layers of rock to a place where it could no longer move up. The trap here was the McMurray sands. As the oil moved into the sands, it began to degrade. The lighter hydrocarbons either evaporated or were consumed by bacteria, leaving behind the thick, viscous bitumen. This bitumen has high amounts of sulfur, an impurity that must be removed before further processing and transport—hence the giant golden pyramids of sulfur I saw north of Fort McMurray.

Fast forward to the most recent Ice Age, from about 2.6 million to 12,000 years ago. As glaciers receded over the sand beds, debris was left behind, obscuring the oil sands. The Athabasca River, flowing through what is now Fort McMurray, cut a channel through the overburden to reveal the vast oil deposit.

It's a funny term, "overburden." It's the same term used in coal mining in the Appalachians, referring to the mountaintops that need to be removed to get to the coal seams. In Alberta, overburden refers to the boreal forest that grew in the glacial debris. In the oil sands mining operations, that's the layer they need to remove to get to the oil. In other contexts, people call it "the lungs of the planet." Recent research suggests that the boreal forests actually store more carbon than tropical forests.[2] Keeping carbon out of the atmosphere is desperately important in controlling global warming. It's all about the framing.

Cutting through the sediments, the Athabasca River exposed the oil sands. First Nations people knew about this oil for a long time. They used bitumen as caulk to make canoes and baskets water-tight. When explorer and fur trader Peter Pond paddled down the Athabasca in 1778, he saw the deposits and wrote of "springs of bitumen that flow along the ground."[3] Like the Bakken, it would take many years and the development of some serious technology before people figured out how to profitably extract the oil from this field.

The Cretaceous period that produced the oil sands had few or none of the vast ice sheets that currently cover the poles. We think of our home as a water planet today, with better than two-thirds covered with ocean. Back then, it was an even more watery place, with sea levels up to 800 feet higher than today, and current shorelines far underwater. The global average temperature was about 97°F; our current global average is 57°F.[4] Fossils of those hadrosaurs from the Cretaceous have been found in Alaska.[5]

Then, in an event that had nothing and everything to do with climate, the Cretaceous came to an end 65.5 million years ago in a couple of hours. An asteroid hit what is now the Yucatan Peninsula at a speed of about 45,000 miles per hour. Computer models suggest that it produced a fiery plume reaching halfway to the moon before collapsing in a pillar of incandescent dust. Fires consumed about 70 percent of the earth's forests, and three-quarters of all species, including the dinosaurs, became extinct. The horseshoe crab, whose eggs are the beloved meal of the red knot, was one of the survivors. The dust from the impact, so thick that it blocked out the warmth of the sun, plunged the earth into a period of cold.

In Bowman, North Dakota, not far from the endpoint of my bike ride, is a geological formation known as Hell Creek. At the time of the asteroid impact, the Hell Creek landscape consisted of steamy, subtropical lowlands and floodplains along the shores of the inland sea. In April 2019, a controversial and flamboyant paleontologist announced that his team had discovered a site there that preserves many of the artifacts of the actual day of the impact.

Robert DePalma has raised a few eyebrows in the paleontology community. He is curator at the Palm Beach Museum of Natural History, a struggling museum with no exhibition space. A graduate student at the University of Kansas, he has yet to obtain his PhD, typically the credential needed to advance in the field. He has posed for Indiana Jones–style photos at the site, complete with vest and bullet-stocked belt.[6] He named the site Tanis, an actual Egyptian lost city portrayed in *Raiders of the Lost Ark*. In fact, DePalma modestly proclaimed that his discovery was "like finding the Holy Grail clutched in the bony fingers of Jimmy Hoffa, sitting on top of the Lost Ark."[7]

Yet after working at the site since 2004, DePalma has accumulated some remarkable evidence, along with endorsements from some of the most prominent paleontologists in the field. He found tektites, tiny pebbles with the chemical signature of the asteroid, at the site, along with the miniature craters they made as they rained from the sky. The surrounding forest had been on fire, as evidenced by the abundance of charred wood and charcoal at the site. His team found fossil fish with tektites lodged in their gills. There was evidence that a huge wave had been generated by the impact. And perhaps most intriguing, the site produced the burrow of a small weasel-like mammal, representative of the mammals, like us, who would inherit the earth.

The prairie that I rode across, between the boreal forest of Alberta and the Bakken of North Dakota, was shaped much more by the retreating glaciers of the Quaternary period, prowling the landscape tens of thousands of years ago, than by the steamy swamps of the Cretaceous. On my bike trip, I rolled through an area of the Great Plains known as the Prairie Pothole Region. The glaciers left shallow wetlands called potholes across the area, which has little drainage. The potholes fill with water in the spring, providing habitat for migrating waterfowl.

On an earlier ride, I had tagged along behind Concetta on a birding visit to a pothole in northern Montana. It was a veritable wonderland: geese, ducks, shorebirds, terns, killdeer. A grebe with two young ones on her back floated by. This was the landscape for much of the ride across the northern Plains, even though many of the potholes have been lost to agricultural development.

Riding south across the Canadian prairie, I was on relatively high ground. Much of the Prairie Pothole region is on a plateau known as the Coteau du Missouri. Were I riding through this part of Saskatchewan thirteen thousand years ago—and had been able to outrun saber-toothed cats and such—I might have caught sight of another, different sea on the far horizon. Unlike the inland sea of the Cretaceous, this one was freshwater and did not extend from the Gulf of Mexico to the Arctic. During the Ice Ages, as the great ice sheet receded, a vast lake formed, easily the largest in the world at the time, far bigger than all of the present Great Lakes combined. At its greatest extent, it covered parts of Saskatchewan, Manitoba, Ontario, North Dakota, and Minnesota. Geologists named it Lake Agassiz, after the Swiss-American naturalist Louis Agassiz, who first posited that much of Europe had once been covered with ice. His namesake lake provided a piece to the jigsaw puzzle of the earth's climate history.

About thirteen thousand years ago, as the earth was in a warmer period of the Ice Ages, the climate experienced an abrupt transition, like flipping a switch, and returned to cold conditions. This sharp cooling, known as the Younger Dryas event, was most apparent in Europe, Greenland, and northern North America. How could climate change so fast? It was a question not at all irrelevant to projecting what the future climate of the planet could look like. One scientist proposed an answer, intimately tied to this vast Ice Age lake.

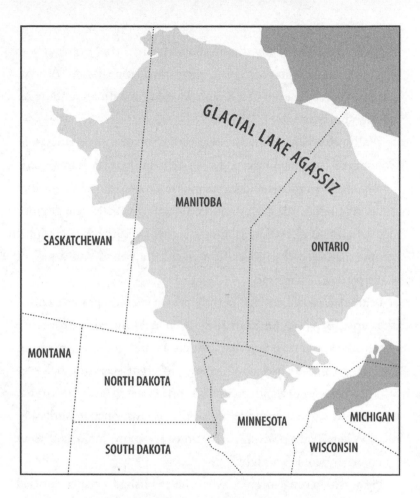

Extent of glacial Lake Agassiz, from Redekop, 2017.[8]

I first heard of him early in my career, at my first meeting of the American Geophysical Union, the great gathering of earth scientists. There was a certain buzz in the paleoclimate section, the people who study the climate history of the planet. Overheard:

"Is Wally here?"

"I wonder what Wally thinks about this . . ."

Most of us wore standard issue, preprinted name badges. A gray-and-red-haired, cherub-faced man walked by with an entourage. His handwritten badge: "Wally."

Wally was Wallace Broecker, professor of geochemistry at Columbia's Lamont-Doherty Earth Observatory. A human idea factory, his forte was generating new, often controversial theses that could set the scientific community ablaze in debate. The popular idea of climate scientists is that they work with vast, complex computer models. His ideas formed with a pencil and a pad of white paper.

I came to know Broecker peripherally. He could produce child-like, epic (or epochal) tantrums when dealing with grown-up things like budgets and grants. But like a child grabbing toys from a chest, he picked tools from many different scientific fields to build sometimes outlandish, always intriguing ideas. Kirk Bryan, a pioneer in computer modeling of the oceans, once remarked to me, "When Wally's wrong, sometimes it's more important than when other people are right."

He wasn't wrong in 1975, when he published a paper entitled "Climatic Change: Are We on the Brink of a Pronounced Global Warming?" In it, he claimed that:

> The exponential rise in the atmospheric carbon dioxide content . . . by early in the next century will have driven the mean planetary temperature beyond the limits experienced during the last 1,000 years. [9]

That's pretty much what's happened. With that paper, Broecker popularized the term "global warming." Later, Broecker would claim that he was just lucky in his prediction, that he was right for the wrong reasons.[10] I'm not convinced. Building on an array of observations of how the earth's climate has changed, he seemed to have an intense and intuitive sense for how the planet works.

Broecker was also fascinated by the ocean's role in climate. In the Atlantic, warm, salty water flows north in the Gulf Stream. As it cools and releases its heat, it warms Europe. At the same time, this now cold, salty water is more dense, sinking to become some of the deepest water in the ocean. It's much like opening your bathroom window after a shower in the winter and feeling the cold air fall by your feet. The sinking of cold water happens in fairly small areas of the North Atlantic. Overall, this flow of warm surface water north and cold, deep water south is known by geeky oceanographers as the "Atlantic Meridional Overturning Circulation." Broecker dubbed it, more memorably, as "The Great Ocean Conveyor Belt."

What if this conveyor belt were slowed or stopped? The warmth brought by the ocean to Europe would drop as well. Perhaps, Broecker thought, this was what happened during this sudden cold interval, the Younger Dryas. But what could possibly put the brakes on this global circulation? His restless curiosity turned to northern North America and Lake Agassiz.

Back thirteen thousand years ago, the world's largest lake was backed up behind an ice dam, looking for a way out. When an ice dam fails, it does so quickly, producing some of the largest floods known on earth. When Lake Agassiz's ice dam broke, it drained and spread fresh water into the ocean, probably within a single year.[11] In my imaginary Ice Age bike ride, I might have been able to watch the lake disappear.

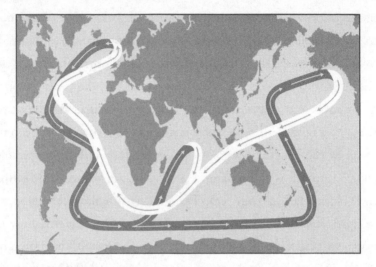

Broecker's Great Ocean Conveyor Belt, a schematic of ocean circulation in which warm, salty surface water moves north up the Atlantic, cools, and forms cold, deep water, whose characteristics can be recognized in much of the global ocean.[12]

Broecker and his colleagues proposed that the fresher water resulting from the draining of the lake poured into the North Atlantic. This less salty water then froze on the surface rather than sinking into the ocean abyss. This effectively capped and shut down the conveyor belt circulation.[13] The giant flywheel ground to a halt, vastly reducing the amount of heat coming north to Europe and triggering the Younger Dryas. That explanation, battled over in the scientific literature for many years, is now largely accepted.

This kind of rapid shutdown of the oceanic conveyor belt is seen as unlikely in the modern world.[14] We don't have giant freshwater lakes ready to burst out as in the Ice Ages. Yet there may be another mechanism in play. Maps of the globe show a record-hot planet almost everywhere except a record-cold North Atlantic. It's speculated that melting of ice from Greenland is producing a blob of cold water that may be acting to slow down the conveyor belt circulation.[15]

Though a face can no longer launch a thousand ships, an idea can launch a few. The climate conveyor belt hypothesis led to explorations and measurements by research vessels on both sides of the Atlantic. From these oceanographic studies, we now know far more about the real ocean conveyor than we did at the time of Broecker's pencil-and-paper cartoons.

Wallace Broecker died in 2019. During his career studying the Earth's history, he was always impressed by how volatile its climate has been, sensitive to small changes and susceptible to major shifts with startling speed. Regarding our emissions of greenhouse gases, he famously said, "The climate system is an angry beast and we are poking it with sticks."

As I approached the US border near Fortuna, North Dakota, I began to notice the pumpjacks, the slowly nodding donkeys drawing oil out of wells. Globally, there are thousands of holes drilled in the ground, punching the ground like a sewing machine, reaching the carbon laid down millions of years ago and bringing it back into the atmosphere. In my head, I think of them like Broecker's sharp sticks.

Fossil fuels have brought a vast increase in our standard of living, back from when pack animals and human backs powered our advances. They enabled my air flight to Fort McMurray and make our house warm in the winter. But ever since Edwin Drake's first oil well in Pennsylvania in 1859, we have been digging up the Cretaceous and making it our own. Not the dinosaur part, and alas, not the Raquel Welch part. Because of the greenhouse effect, all that carbon, in the form of carbon dioxide in the atmosphere, is making the planet steadily warmer, as we set off on the path to the climate of the Cretaceous.

As I was writing this at my home in Maryland, Concetta went out on the front porch in the evening during one of the ever-more-frequent heat waves to come across the central and eastern United States. This was the third day of over 100°F.

"Wow, sure is swampy," she said.

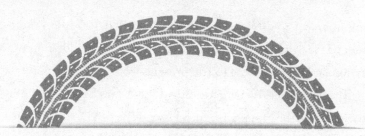

12

MEET ME IN MOOSE JAW

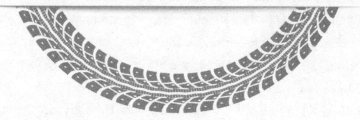

Time to get back out into the wind. Fifteen miles south of Chamberlain, Saskatchewan, I pulled the bike over alongside a wheat field. I was nearly spent. Fortunately for me, the torrent I was riding in was just a crosswind. I was working hard, but I only had to worry about keeping the bike from being blown over. I couldn't imagine how anyone could be riding directly into that wind. Then again, this was Lynn we were talking about.

The day had been planned for the better part of a year. At a book reading I was doing in Bethesda, Maryland, the previous summer, I noticed several women in cycling attire in the audience, members of a Virginia-based group called Babes on Bikes. It turned out that a full-fledged cycle gang had ridden across the Potomac River to come to the presentation.

The leader of this energetic crowd was Lynn Salvo. At the end of the reading, the group introduced themselves. All but Lynn headed back on the twenty-five-mile ride across the Potomac. As we were packing up, Concetta and I invited Lynn to join us for a late lunch. We traded stories of our cross-country rides. Her next endeavor would prove to be more formidable yet. She was planning to try for the equivalent Guinness Record across Canada that next summer. Funny, I was aiming to ride across Canada too, from north to south. Lynn proposed that we should ride together for a while. Our routes intersected in south-central Saskatchewan in a city where, not surprisingly, the town symbol is a giant moose statue. "Meet Me in Moose Jaw" had a nice ring to it.

It wouldn't be our last ride together. The following year, we would put in nearly a thousand miles together on the East Coast, so we would get to know each other pretty well. Before the Moose Jaw ride, I stopped by her house in Virginia to go over endless logistics: routes and motels and traffic and shoulders, guessing where roads turned to gravel, trying to anticipate the miles and the climbs and the wind. Her husband, Giovanni, wandered in as we pored over maps and spreadsheets.

"You're going to have a hard time keeping up with her," he observed.

"I'm beginning to get that idea," I nodded.

We both had our reasons for the rides, our causes. I wanted to go to the places where the seeds of climate change are sown.

Lynn's ambition was different: She wanted to imprint on North America, with her bicycle track, a peace sign the size of the continent. She was doing this in memory of her brother, a pilot lost in the Vietnam War.

"If I can possibly keep one family from going through what we went through, it will be worth it," she said.

"How did it happen?" I asked.

"You know, I don't know just exactly what caused the crash. I understand that you can find the details easily enough. We just felt the repercussions in the family. His son was born three months after he went down. They never met each other."

Captain John Thomas West, United States Air Force, went missing in action over Laos on January 2, 1970, after his F-4 Phantom crashed. His wingman never saw a parachute. He has quite a presence on the web. He is strikingly handsome on the beach at Cam Ranh Bay in Vietnam, forever twenty-eight. Five pages of remembrances follow, some from people who never knew him but wore his memorial MIA bracelet.

I came to be part of Lynn's grand peace sign, meeting in Moose Jaw to ride for a day eastbound to Regina, Saskatchewan. I also had a less altruistic incentive. Lynn's friend Susie Schmitt was helping with logistics and driving the car carrying her equipment. She was support-and-gear, or SAG in the lingo. I had been riding fully loaded, complete with clothes and tools and camera and laptop. For me, the prospect of riding thirty pounds lighter was positively delicious.

The choreography needed to be perfect, though. Lynn would have precisely twenty-three days to ride east from the Pacific coast of British Columbia, and I needed to ride thirteen days southeast from Fort McMurray, Alberta, to meet on July 9. The plan was held together by 124 gossamer spokes and a handful of thin tendons. A parting of any of them would leave our rendezvous by the side of the road.

It was a near thing on my end. South of Fort McMurray in the boreal forest, I lost a day to horizontal rain and 40°F temperatures. Returning to the warm and dry motel room, I frantically replotted my route to straighten it out and regain the time I'd lost. As I hit the prairie winds of Saskatchewan, I was on track for Moose Jaw.

On July 9, I wound my way through the neighborhoods of Moose Jaw and found the home of Glenda, our host for the night. The town was a tree-lined oasis in a sea of grain fields. I was alone and early, so I took advantage of the shade to relax and blow through the rest of my water. Shadows were growing long, but the east wind had not ebbed in the slightest. As I waited on the front steps, I saw a cyclist moving slowly up the street. Lynn rolled into the driveway and stiffly kicked her leg over the seat. We hugged, but Lynn was a little blank. I could read the exhaustion in her eyes, an eleven-hour struggle against the wind bringing a certain emptiness. She had ridden 82 miles dead into the same wind that had been doing its best to blow me off the road. Later she would say that it was one of the hardest days of her ride across Canada.

Food, water, and electrolytes are wonderful things. Lynn and I recovered in time for a selfie with Susie at the Moose Jaw moose, who incidentally is named Mac. He was later at the center of an international incident when the town of Stor-Elvdal, Norway, provocatively erected a stainless-steel moose that it claimed to be the world's largest at thirty-three feet, a foot taller than Mac. Diplomats reached a solution, however, with an antler extension for Mac and the acknowledgement that Stor-Elvdal's would "forevermore be known as the shiniest and most attractive moose in the world."[1]

It was saskatoon season, a local fruit a little bigger than a blueberry. The purple fruit is intensely tart and sweet. Glenda's breakfast was a magical saskatoon crisp, which we lingered over because our ride that next morning was planned as an easy, flat fifty miles from Moose Jaw to Regina, the provincial capital. We rolled out of the sleepy city and onto the Trans-Canada Highway.

Back on the plains, I settled into the standard rhythm of riding with someone else. Lynn took the lead at a steady pace. I assumed that we would switch off in front, since wind resistance in the lead position makes riding there noticeably harder. Lynn quickly stopped me from moving up. The Guinness standards were quite clear that potential record holders could not use others to help them, though SAG vehicles were allowed. For me, it was a shrug and a motto for the day: Ain't too proud to draft.

This was my first day following the gray braid that came out of the back of her helmet. She doesn't carry a spare ounce on her small frame, and I discovered one of our many differences. She's an athlete; I'm a schlub. Originally she spent some years doing triathlons, only later falling in love with distance cycling when body parts failed: two rotator cuff surgeries put an end to her swimming. But the genesis of her days as an athlete came earlier, when she was more than a little irritated at the prospect of turning fifty.

"That year I started my PhD, started a business, and started running, all at once," she told me.

That was nearly twenty years ago. She did them all. Back when she was in college, women weren't welcome to participate in marathons. At fifty-two, she completed one. As a mathematics educator, she started a company that ran summer math camps before selling the business and taking on the transcontinental rides.

"I like big projects. Don't ask me to clean the house. That's way too hard."

Her gear is completely tricked out. Lynn's mechanic assembled her wheels by hand. Aero bars, a handlebar rig that allows a cyclist to assume a low profile, sit atop a carbon-fiber bike. On the road, she gave me a wind shadow—that quiet space behind a wind obstruction—but not much of one. At the front of the bike is a Garmin navigation system over a mount for a video camera. She has all kinds of other gear in the support car. At one point during our later rides, I needed a new tire. From the motel, I put in an order to a Denver warehouse for overnight delivery. But, of course, Lynn had one in her car's wheel well.

Lynn is relentless, but not quite fearless. Big electrical storms happen out on summer roads, and she's more than a little jittery about them. Halfway into the ride to Regina, thunderheads were building. From the Doppler, it looked like we could ride out of it and they would sweep behind us. But the crackling was up there in a darkening sky.

Lynn yelled out, only half-joking, "If I get struck by lightning, tell Giovanni I love him."

"And I'll mention that your last words were 'Cut Dave into the will,'" I added. Not long after, the storms scooted off to the south.

Despite all her gear, I'm a bit better than Lynn on mechanics. I have a certain pride in being able to fix things by the side of the road. Lynn is seriously organized, with a checklist each night, meticulous in how she rides and how she prepares. She's fixated on her equipment, that sweet Cannondale carbon-fiber machine that would take her to the coast of Nova Scotia. She cleans the chain every other night in the room and wipes the mud off the frame. Nobody touches the bike but her and her mechanic. Both Lynn and her bike are well turned-out. My bike is old, heavy steel, with that rust spot on the frame where the sweat drips from my chin. It's dirty, falls a lot, and specializes in getting back up. I've ridden 200 miles with a bubble on the tire and 500 with a cracked rib. Lynn and her bike don't fall.

On a later ride in New Jersey, we were near the end of a hard, 60-mile day. We ducked under a highway overpass on hearing that familiar crackle of thunder, now a little too close. I checked the Doppler and quickly announced that we were going nowhere until that big red-orange blob passed overhead. The tempest revealed itself out of dark black clouds and threw sheets of rain all around our cozy perch. After the sun broke out, we hacked out the last ten miles on a mud-puddle trail. With the rain, we had neglected to stock up on drinking water for the afternoon, and Lynn was out. I gave her half of my last half-bottle. She used it to wash the grit off her brake pads.

Considering all of our differences, Lynn and I came to give each other a lot of space. Good fences make good neighbors, and a wide berth holds true when on wheels. We've both done transcontinental rides, and mansplaining distance cycling techniques to someone who's ridden tens of thousands of miles wouldn't go over well. We claim that we're always ready to learn, but in reality we're both pretty set in our cycling ways, not to say pigheaded. For example, Lynn rides with a rear-view mirror, and I don't. I want to keep my focus ahead, on that next pothole that I have to dodge. Then I simply ride like there's always a semi on my elbow. Lynn believes that my mortal soul is in peril without a mirror and gnaws at me during our rides to that effect. She may yet prevail.

Late in the afternoon the skyline of Regina began to appear over the prairie. The notion of riding without bags and with a support car became more and more appealing as we approached the city where we'd part. Susie was amazingly hardworking and loyal, waiting in Lynn's "tin can" so she could shoot video and showing up with ice water in the hottest part of the afternoon.

I set about making a plan for stealing Susie, for having her abandon Lynn and follow me down to the Dakotas. It would be a modified Bonnie-and-Clyde romance: She could rob banks during the day while I rode. I laid on my very best Jerry Lee Lewis: "Hey baby, stick with me. I'll treat you right."

Alas, it was to no avail. That next morning Lynn and Susie disappeared into the eastern prairie, and the road got lonely. It made for odd conversations. I spent the first hour talking with my Trek, trying to sound convincing that I hadn't been making eyes at Lynn's fancy Cannondale. Grudgingly, the faithful Trek came around, and it's a good thing it did. The crosswind howled again, and the day was another serious slog. Sometime during the morning, I passed within ten miles of Avonlea, Al and Randi's town. I'd be back on this road the following spring.

I took a breath by the side of the road. The crosswind that was battering me had backed completely around from the west. This day it was behind Lynn rather than in her teeth, and it blew her across the prairie for an incredible 139 miles. The same blow would be my crosswind riding south. Going upwind would have been unrideable, but even with the crosswind I faced a serious battle for the last 40 miles into Weyburn.

That morning we had ridden off on different journeys, both quixotic. Lynn would use the continent as her canvas, visiting peace memorials across the rest of Canada in the process of etching her peace sign on the earth. I would go on seeking some clues to the nature of this industry at once so tightly woven into modern life and so threatening to it. At least we both had some tanned, lean legs.

Looking back, I've come to believe that Lynn is one of the strongest people I've ever met. It's forced me to think about what I mean by "strong." It's surely not being the fastest or the most powerful. There have certainly been times when she's turned to me when climbing and said, "That's all I've got. Please feel free to pass."

Her strength is in a certain quiet perseverance, in how she can ride thirty-two days consecutively without a rest day. She believes that it is still possible to be an athlete at seventy. I speculated that maybe the monumental rides were her way to rage against the dying of the light. Lynn would have none of it. She replied, "I don't feel like I'm raging. I feel like I am embracing every ounce, minute, inch of life I have left."

When I came back home to Maryland, I went on another ride, this one down a quiet stretch of the Potomac. The Capital Crescent

Trail is a pathway through the woods, so folded up in forest that you can scarcely believe you're approaching a city. Eventually the bustle of Georgetown made itself known, and I rolled past the Watergate and the Kennedy Center. On a rise, familiar white columns. I turned left at the Lincoln Memorial and parked the bike.

I looked down toward the deep black scar of the Vietnam Veterans Memorial. It's always striking how the polished marble reflects the images of the pilgrims, coming from every corner of the country. And how many names there are. I took a deep breath and walked down to meet John T. West, Lynn's brother, at panel 15W, line 117.

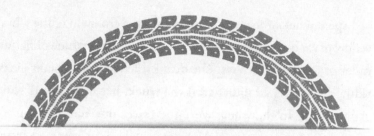

13

THE CROSSING

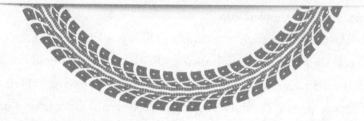

There's a standard entry in a ship's log when taking over the watch: *Underway as before*. The bike and I were underway again, heading out of Regina, but without the company. The loneliness shouldn't have hit me so hard. I was better than 800 miles in, most of it solo. I was used to taking on bright sun, blasting wind, and the huge swings of temperature on the prairie. But now I was back to my own thoughts all day, feeling very small under a very big sky, with few options if I simply gave in to exhaustion and knocked on the first door I saw. That farmhouse oasis on the horizon could be either abandoned or home to a vicious German shepherd.

I spent much of that day leaning into the crosswind. The vibrant yellow of canola fields gradually gave way to the soft blue of flax and miles of windswept wheat. The dust kicked up by a distant pickup riding beside a field didn't trail the truck, but instead was blown right over it. In the wheat were the traces of a hundred gusts of wind sweeping across from the horizon, eddies in a massive stream flowing across the plains. I was dead tired that day, but riding by the wheat field, I had a certain thought: I got this. Despite aching knees and feet, clenched fingers and shoulders, my bike and body were pretty well locked in, sinews standing out in my legs. I looked out across a vast prairie and a big sky and realized what a beautiful place I had stumbled into.

Posted on the edge of a canola field was a lovingly constructed metal sign with nothing but fields behind. The sign held cut and painted images of a white building framed by swing sets and a flagpole. It read "Polish Draw School 1906–1961." Once upon a time, these vast fields were home to running, playing children and a thriving community. But up and down the Great Plains, farming needs many fewer farmers than a century ago, and the schools and swing sets have disappeared with them.

Not far away was a field full of stacked pipe. The pipe was in storage for the construction of Enbridge Line 3, which is planned to go east from Hardisty, across the prairie provinces to Minnesota. In every direction, oil is trying to get out of the oil sands, and carbon out of the ground.

Battered by the wind, I rolled into town at day's end under a dramatic sky of racing, dark-bellied cumulus. A bright LED sign announced Weyburn, the Opportunity City, with an image of pumpjacks—the nodding donkeys, the thirsty birds ubiquitous in the oil patch. I had dinner at the nearly empty hotel restaurant, then collapsed facedown in bed by eight. The real pumpjacks wouldn't be far away.

In planning the route, I thought about the way south from Weyburn for a long time. The path south across the prairie to this point featured an unbroken string of small-town motels spaced at 50- to 70-mile intervals to the end of the journey. This spacing was an important consideration. I had no camping gear, and more than 70 miles would be a stretch for old legs.

My attention focused on the 142-mile stretch between Weyburn and Williston, North Dakota, where there was one and only one place to stay: The Old School. Set in the town of Fortuna, North Dakota (population twenty-one), The Old School is exactly that, an abandoned elementary school converted to a motel of sorts. The ad for their bar features a cartoon of a leggy young lady sitting on a desk peering over glasses, looking like the teacher ten-year-old boys remember long after school days are done. The bar, of course, is known as the Teacher's Lounge. The notion of a cold beer in the good old USA put wings on my feet.

Early the next morning, I made my break for the border. The ride was due south on a crisp, clear morning, to where the blue flax fields gradually become interspersed with oil tanks. The soundtrack was the rattling of a stiff wind superimposed on the creaking of pumpjacks. On the map it looked like there wasn't much between Weyburn and the border, only a town called Tribune. Approaching the crest of a hill where Tribune was supposed to be, I could see just a few structures coming into view. Only one had the appearance of a store, so I stopped in.

Dallas Locken runs the combination coffee shop and insurance agency in Tribune. It was the only place to stop for the entire fifty-mile day. He gave me a Coke on the house, and we talked. I mentioned Tribune, Kansas, a town that I'd ridden through on an earlier journey. Dallas had been there, and across all of the plains in between. Out on the prairie, it seems that people on either side of the border are closer to each other than to distant cities of the east and west.

Insurance people have an interesting take on climate, because they tend to take the long view. Most of us find it hard to look past the next weekend or the next quarterly earnings statement, but that doesn't work in insurance. Dallas gave his perspective:

> Insurance policies used to be priced based on crime, and so rural insurance was cheap. Now all the companies talk about are flood areas. This part of Saskatchewan used to be much drier, only about seven inches of rain a year. Since 2009, that's about doubled, and the rain tends to come in big downpours. Places that have never flooded now do. The "once-in-a-hundred-year" floods come much more often.

Data from Environment Canada bears him out on increased rain, though not quite a doubling.[1] He continued:

> On the other hand, with longer growing seasons and more rain, there's more land put into crops than hay,

and a lot more money to be made. Land prices have been going through the roof. The big operators are driving it. One farmer I know was retiring and hoping to get two million for his land. An investor from Calgary offered him three million, sight unseen. The big guys buy up huge swaths of land and then have outside crews farm it. Good for the retiring farmer, but terrible for the young farmers trying to break in. They just can't afford the land. There are no small farms or farmers any more.

Climate is changing, and people and the land are responding, not always in predictable ways. Dallas filled up my water bottles, and I rolled out onto the 17 miles between the United States and me.

I flew across the green prairie on the CanAm Highway, counting down the miles. On the horizon appeared a copse of trees, then low buildings. I coasted past Canadian Customs and stopped before what appeared to be a deserted US Customs station. An emphatic voice over the loudspeaker at my elbow made me jump.

"Step back from the white line."

Welcome back. A brief conversation with the agent at the window, a glance at the passport, and I crossed into my home country in bright afternoon sun and long shadows. The sign on the grassy, rolling hills read WELCOME TO NORTH DAKOTA—LEGENDARY. There's something different about coming into Dakota from the Canadian prairie. Maybe it's just a little more isolated, a little more rolling, a little more extreme. Kathleen Norris, in *Dakota: A Spiritual Geography*, described the country as "a terrifying but beautiful landscape in which we are at the mercy of the unexpected, and even angels proceed at their own risk."[2]

Eight anxious miles passed without a structure in sight. I was 50 miles in, with lengthening shadows. Imagining that a town was up ahead seemed a leap of faith. Then the spire of a chapel emerged from the horizon. Soon after, the classic shape of an elementary school appeared, familiar except for the beer ads on the walls. And most schools don't have TEACHER'S LOUNGE in big letters on the side. This was the place.

Well-infused with road dust, I wandered into the bar. It was roughly where the front office would have been, and the adjacent library had been taken over by a pool room. Behind the bar, bottles were lined up on bookshelves. A couple of locals were nursing beers and commiserating with Carol, the bartender. She was wiping down the bar, shaking her head and grumbling about the last customer.

"They were giving me grief about how much a six pack is. I *know* how much a six pack is." She turned to me. "Can I help you?"

"I'm hoping you have a room for me."

"Sure do. Right down that hall. It'll be nice and quiet. Bath and showers on the left."

The hallway was universal elementary-school style, linoleum flanked by glazed cinder block walls, but the classrooms had been partitioned off into motel rooms. The showers kept a nod to their former life: a brass "Boys" nameplate remained on the door.

Back in the bar, 50 miles of fevered dreams climaxed in a single exquisite beer. Then I ordered pretty much everything on the menu. The rattle of a grease explosion burst from the back, sounding like my whole frozen dinner splashed into the deep fat fryer at the same time. It would hit my stomach like a brick, but then again, I had a brick's worth of room.

Needing to walk off some of the Teacher's Lounge cuisine, I wandered across the road to the streets of Fortuna. I came across an abandoned church, what's known on the Plains as a prairie light-house. The chapel was soundless but for the flap of birds nesting in the belfry. The windows were boarded, and scrub trees huddled up alongside out of the wind. No one seemed to know its name. I have a fondness for these outposts. On another journey, on the Rosebud Reservation in South Dakota, I had come across another abandoned chapel on a ridgeline on Memorial Day. Flags were lashed to sticks over the graves of three generations of Lakota veterans. These light-houses watch over the land-sea, providing the gift of a first sighting over the horizon and a welcome for travelers and prairie schooners.

Back at the Teacher's Lounge, I needed one more beer for hydra-tion, pausing by the big screen to catch the score on my team, the Yankees. Two men in ball caps and stained jeans were at the table, by looks related. The older one signaled for my attention.

"Want to watch? Have a seat." I did.

"You gents actually Yankees fans?"

"Hate 'em to death. But they knocked off Cleveland in the playoffs last year, so they're good for something."

I negotiated. "Well, my wife's from the Bronx. When we talked with the priest before the wedding, he said we had to raise the kids Catholic and Yankees fans. So here I am."

He nodded. "At least you come by it honestly. Name's Paul, and this is my son Chris. We're oil-field trash." People who get their hands dirty.

"I've been there," I said. "Spent some time on the underside of generators out on the rigs in offshore Louisiana back when. No way Daddy would let me go out on a date with his daughter once he found out I was a roustabout. What do you folks do?"

"See all those wells out there? Lot of machinery. Sometimes it works, sometimes it doesn't. Sometimes it busts above ground, which is easy, and then sometimes it's down in the well, which gets tricky."

"You folks been busy?"

"Plenty of rocking horses around these parts, and sooner or later they all need work. But the rig count's still way from the boom back in '11 and '12. Anyway, we work out of Glendive, over in Montana. What brings you to this part of the world?"

"I'm a retired climate scientist, wandering across the Plains."

Paul shifted in his chair. "Climate's always changed. We've had ice ages right over here before."

"Absolutely. But climate's changed with carbon dioxide for at least the last 800,000 years, including in the ice ages. And now our carbon dioxide levels are off the stops, way more than they've been over that time."

Paul wasn't particularly swayed. "Climate's always changed" is probably the most common argument to make the case that people have nothing to do with climate change. There's no question that the planet is warming and the big modern increase in carbon dioxide is coming primarily from burning fossil fuels. The undertone from people in oil and gas is essentially: Are you really sure this isn't just a natural thing? I really am sure, as are the vast majority of climate scientists. But as Upton Sinclair once said, "It is difficult to get a man to understand something when his salary depends upon his not understanding it." Paul and I switched back to discussing the Yankees bullpen.

Paul and Chris finished up with handshakes all around and headed for their truck. I tottered back to my room. There was no front desk, and I'd be on my own in the motel that night. The echoes of books opening, chalk scratching the board, and children's laughter were still around the rooms. The Teacher's Lounge and its liquor supply were safely behind a steel mesh screen.

The next morning Debbie, behind the counter at The Old School's convenience store, told me it was going to be a rough one. Mid-nineties, a headwind, and nothing between there and Williston. "Well there's Dan's place about halfway. Dan might be open. But you can never be sure." (Dan turned out to be thoroughly closed.) I set out with three bottles of water and two of Gatorade. That was a lot of weight. But it wasn't enough.

The only way into Williston involved a ride on US 85. In Maya Rao's book on the Bakken, *Great American Outpost*, she writes, "Highway 85 was the most dangerous route in the oilfield, and everyone seemed to know at least one motorist who had died in a wreck there."[3] I was a little edgy about the day's ride. But the rig count, a measure of the amount of drilling activity in the area, was way down from the year her book was written, so the traffic shouldn't have been too bad. That was what I kept telling myself.

Rolling out of Fortuna in the early morning with the sun to my back, I received a salute from a flickertail, more properly a Richardson ground squirrel. North Dakota's official nickname is The Flickertail State. He stood on the edge of the high grass, cheeped a sound something between a laugh and a bark, and then dove back into the burrow.

The unanticipated proved to be the biggest problem: the shoulder. Only bad choices presented themselves riding this stretch of US 85, a cycling Scylla and Charybdis. On the left were deep rumble strips. If I wandered in at speed, they would rock my world, and possibly my rims. On the right, soft gravel. Going in fast on that side would mean I was going down. The bike would probably be okay, but I'd spend some quality roadside time picking pebbles out of my skin. In between was barely six inches of angled pavement. And beyond the rumble strips was the domain of the oil trucks. If it's any consolation, I'd never know what hit me. The only saving grace was relatively light traffic and long sight lines. I kept an iron focus on staying in those six inches, sneaking out occasionally into the open road when the coast was clear for a few minutes.

The remoteness of the country held its own set of hazards. I passed a sign for saltwater disposal, another euphemism of the oil field. The process of fracking leads to a backflow of brine, remnants of the ancient sea that laid down the oil deposits. The brines also contain drilling fluids, toxic chemicals, sand, sediment, and hydro-carbons. In the oil patch they call it "bile of the subterranean."[4] This waste is injected into deep wells, if it makes it there. In 2015, right off Route 85 at Blacktail Creek was the largest brine spill in the nation to date, at roughly three million gallons. It's far from the only one. North Dakota has had at least 868 brine spills since 2008.[5] These spills are actually worse than oil spills, because it can take years for the salt to leach out of contaminated ground-water. The land might as well be a desert.[6]

I also passed not far from the site of the curious case of the radioactive socks. All that sediment that emerges in the backflow from fracked wells needs to be filtered out so the waste can be transported. The filter socks used in the process concentrate naturally occurring radioactivity. They can get seriously hot, presenting one more disposal problem. Thus, it came to pass, in 2014, that an abandoned gas station in the remote town of Noonan was discovered to contain roughly two hundred industrial-sized garbage bags of radioactive filter socks. Windy, sparsely populated Divide County, hard by the Canadian border, now had its own toxic waste site.[7]

Within 10 miles of Williston, both oil traffic and drilling activity increased noticeably. In the heat of the day, headwinds were firing up. I settled into life in the blast lane. I started metering my water, drinking only every 5 miles and riding dry. The road offered no place to sit. Horizontal surfaces were either the needles of dry grass stubble or black, baking asphalt.

I stopped by the side of the road, getting off the bike for a water break by a sign for the inauspiciously named town of Bonetraill. It was 95°F, and water was getting tight. Besides a slowly turning pumpjack and a cell tower, little sign of humanity presented itself. Off in the distance oil well flares flickered languidly on the horizon, a touch of Mordor out on the prairie.

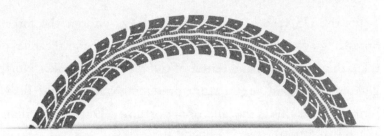

14

WILLISTON: BOOMTOWN, USA

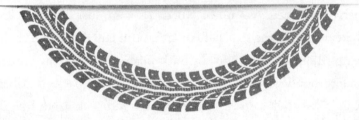

The plains north of Williston shimmered in the heat. Occasional trees huddled down in the draws, but mostly wide-open space stretched to the horizon. The road widened from two lanes to four as traffic picked up dramatically. All heavy equipment—pickups, semis, oil field equipment trucks—passing within a few feet, sometimes inches. I could feel my shoulders tense up and my grip on the handlebars tighten. This was no place to make a mistake. The billboard from the tourism board made the announcement: WELCOME TO WILLISTON. BOOMTOWN, USA.

It's the US version of Fort McMurray, without the forest. Sprawling out on the prairie, Williston is the unofficial capital of the Bakken oil fields and center of the region that makes North Dakota the second-largest crude producing state, behind Texas. In many ways this is the land of opportunity. During the financial crisis of 2008–9, the national news shows showed a splash of low unemployment in this remote corner of the High Plains. Oil field workers lived in RVs in the Walmart parking lot over a North Dakota winter, reminiscent of the Fort McMurray tool sheds, winter quarters for many during the early oil sands surge.

Williston's boom didn't happen overnight. For decades, geologists knew there was oil in North Dakota, just like they knew there was oil in the sands of Alberta, both laid down by a shallow sea millions of years ago. In both places, it wasn't in immense underground pools. As my friend Bryan said back in Alberta, "Can't just stick a straw in the ground and suck it out like they do in Saudi Arabia." In western North Dakota, the oil was locked up in shale formations, in particular, the shale formation deep underground known as the Bakken. "Tight oil," as they say in the business.

Petroleum engineers knew that the key to releasing shale oil and gas was pulverizing the shale, using the process known as fracturing. Oil and gas would then flow through the cracks. In 1967, the Atomic Energy Commission even detonated an atomic bomb underground in a New Mexico shale bed for this purpose, dubbing it Project Gasbuggy. It worked, sort of. A vast amount of gas was released. The bad news, however, was that it was all radioactive.

More technology was required to access tight oil and gas. One necessary development was horizontal drilling, which is the ability to drill down to a particular layer of shale, then turn and drill directly into that layer. Then, rather than using an atomic bomb, high-pressure fluid injected into the well creates micro-cracks in the shale. The drilling fluid contains sand to keep the cracks open, and lubricants, some quite toxic, to prevent clogs and bacterial growth. This fluid is a major component of the brines disposed of, and often spilled, in western North Dakota. The overall process is hydraulic fracturing, or fracking. Early on, "fracking" was seen as a pejorative shorthand. Now it's touted on billboards all over Williston.

Fracking largely came about because of the persistence of one already successful Texas wildcat driller: George P. Mitchell. While Mitchell didn't invent fracking, his tenacity in pursuing the technology made it practical. In the later part of his career Mitchell became obsessed with developing the Barnett Shale, a formation 70 miles from Dallas and a mile and a half underground. After years of repeated experimentation, Mitchell's company hit on the right combination of drilling, fracturing, and fluids to profitably bring shale oil and gas to the surface. The result was revolutionary, and it upended the global energy system. Hydraulic fracturing has led to the United States passing Saudi Arabia and Russia as the largest oil producer in the world. Following the oil embargoes of the 1970s, securing an offshore oil supply was once a major driver of US foreign policy. Fracking changed all that.

But in later years, Mitchell was caught off-guard by the reaction to the environmental issues that arose with fracking. In 2012, the year before he died, he co-wrote, with Michael Bloomberg, an op-ed for the *Washington Post* arguing for increased regulation of fracking. "The rapid expansion of fracking has invited legitimate concerns about its impact on water, air and climate—concerns that the industry has attempted to gloss over," they wrote. Mitchell expressed himself more succinctly to his son-in-law Perry Lorenz. "These damn cowboys will wreck the world in order to get an extra one percent [of profit]," Mitchell said. "You got to sit on them."[1] North Dakota, among other places, would prove his concern prescient.

I rolled into Williston completely out of water, dehydrated, and a little dizzy. I stumbled into the first fast food joint, almost passing out after hitting the A/C, then filling and filling and filling my soda cup. The sweat was caked on my face, and it took ten minutes before I realized that I still had my sunglasses on. I started to look out at the world in unfiltered sunlight.

Williston's main drag was like any of a thousand other Miracle Miles in medium-sized towns. Bars, chain restaurants, motels, and strip malls lined the road. That Walmart where the RVs used to park was on the right. It wasn't so different from the strip in my hometown in Maryland, but with fewer trees and a whole lot more pickups.

Coming into Williston was a bit of a shock to me. After weeks of riding through some nearly vacant parts of the plains in both Canada and the States, I was suddenly surrounded by giant machines whizzing by my elbow, and by people. I wondered if some of those people might not necessarily be all that friendly to some guy from the East Coast on a bike. They live in a world very different than mine. I might have had some justified concern about the machines, but not the people. They were unfailingly open, friendly, and willing to help a stranger on his way.

Rolling into town, I was more than a decade behind the original deployment of fracking in this part of the world. At the opening of the millennium, the emerging technology created, in the curious parlance of the oil patch, a new "play" in western North Dakota. In 2004, Harold Hamm's Continental Resources brought in the first commercially viable fracked well in the state. Two years later an EOG Resources well produced oil under so much pressure that the company had to shut down the well until a second one could be drilled to relieve the pressure and prevent an old-time gusher.[2] The boom was on.

All around Williston, a patch of light larger than metropolitan Chicago began to grow on the surface of the Earth, easily visible from satellite imagery, though no one noticed at first. The burning of methane gas from thousands of wells, known as flaring, lit up the land. Down on the ground, truck drivers would park near the flares during the bitter Dakota winters to stay warm. Meanwhile, the boom attracted people, especially during the hardest time of the Great Recession. They were mostly young and mostly male, looking for work, a new start, trouble, or all three. Things got rough. It's not a new story. During the California Gold Rush, one reporter described the scene in 1849:

> The community was composed of isolated individuals, each quite regardless of the good opinion of his neighbors; and, the outside pressure of society being removed, men assumed their natural shape, and showed what

they really were, following their unchecked impulses and inclinations.[3]

As the boom came, crime shot up in sleepy Williston. Maya Rao quotes one roughneck in town: "This place is for people who messed up their lives."[4] It wasn't that different than when I worked on the offshore oil rigs in Louisiana some years ago. Like North Dakota, the rigs were great places to disappear if you were on the run. In trying to make casual conversation with one of the other Louisiana roughnecks, he mentioned to me that he'd done time for "B and E." I racked my brain: *B and E, B and E, B and E . . .* then the light went on: *Breaking and Entering!* It's all about the lingo.

Fourteen years after the initial boom, I thought that perhaps things were settling down in the Bakken. At the airport in Bismarck, the town where I eventually departed, the gate agent asked about where I had been on the ride. As I fumbled to get my bags on the scale, I mentioned Williston.

"I guess it used to be sort of a Wild West town," I said.

She shook her head. "Not just used to be. I drove delivery into Williston not so long ago and trust me, the airport's a whole lot safer. Lot of guys come into Williston, work in the oil fields two weeks, and are looking to party. I would go in with the windows rolled up and the doors locked. A woman on my own, I drove fast, got in, and got out. I know that lots of the roughnecks are good people, getting a paycheck to support a family back home. But there's a lot of trouble there too. I could tell you stories."

The town didn't seem like an easy place for women. I stopped into Williston Brewing Company for a beer, entering using the door handles shaped like repeating rifles. On the way to the necessary stop, the restroom names pulled me up short: "Trucks" and "Garages." Later, on the way back down the strip, a paper taped to the window of a Williston motel gave me a little shiver. It was about a missing person, a woman with a Native American surname. "Last seen leaving the bar in a blue pickup."

The Bakken can be hard on the oil field workers too. It's dirty, dangerous work. Breakfast bars at chain motels offer quick glimpses into other travelers' lives. While waiting for waffles, I asked a middle-aged man what brought him to Williston. He gave a winsome smile.

"I spent my career in the concrete business, and that can be pretty hard on your body. But my son got busted up working out on a rig. They're fixing him up with a big back brace now. I'm wheeling him out of the hospital, packing up his apartment, and we're heading back down to Texas."

Williston's not all strip malls, bars, and oil field offices. I wandered into Rib Fest, a town fair on the grounds of the local university. The school was in the middle of a quiet residential neighborhood, with planted trees, shade, and side streets. At the fair, rig service companies were cooking barbecue and handing out T-shirts. The scene was bands and beer on a sunny Sunday afternoon. People wandered by in random shirts that read things like:

[Pregnant woman]: You're Kickin' Me Smalls

[Goateed man with a lean face]: Please Lord Give Me One More Oil Boom. I Promise Not to Blow It This Time.

[Older man in ball cap, sunglasses]: Take Your Gun Laws and Shove Them Straight Up Your [Democratic logo]

[Very large guy holding very small baby]: Straight Outta Tonga

The last one caught my eye. I realized that I was sitting next to the Nation. In Blair Briody's book *The New Wild West*, she writes of a group of Polynesians who had become a legend of the Bakken.[5] All originally from Tonga, they came from Salt Lake City or Spokane or Monterey. Family connections led them to jobs in the oil patch. They work in a specialty called coil tubing, a process of running bendable pipe down wells after they had been fracked to get the well to start producing oil. The Tongans became known as "the ghost crew," because they spent so little time at the equipment yard and so much time out on the job. They called themselves the Nation. The Tongans tended to hire family and other Tongan friends. Judging by the baby, it looked like some of the Nation were settling down. It made for several picnic tables full of very big people.

I relaxed with my beer to catch the Jensen Sisters on the Rib Fest main stage. One of the sisters wore an evangelical Slay Like David T-shirt. She opened their set talking about their connection to the town.

"We're from Minnesota, but this feels like home. Our dad used to work here all the time."

Williston is one of those places where people come to work, most from a long way away. It's where the money is, but it isn't easy on a family. I remember the advice from a roughneck in the Louisiana offshore who'd come down from Arkansas.

"Always come back home through the back door. That's where he'll be running out. And make sure you check that toilet seat first thing."

The oil patch is how whole swaths of the country keep their families afloat. The notion of working a long way from home is nothing new. Consider the waves of immigrants from Ireland and Italy in the nineteenth century who came to America, sent money back for the family, then eventually sent for the family. Maybe that's what I was seeing with the Tongans.

Besides the Tongans at the Rib Fest, there's additional evidence that people are coming with families. A new high school has gone up in Watford City, the next town south from Williston.[6] This part of North Dakota is going through the same kind of normalization that took place in Fort McMurray after the initial boom, or, for that matter, in Sacramento after the Gold Rush.

The boom times that had brought so many people to Williston ended in the oil bust of 2014. Oil dropped from a high of $108 a barrel in June of that year to $29 by the start of 2016.[7] In many ways, both Fort McMurray and Williston were victims of their own success. By 2014, so much new oil was coming on the market that Canada and the United States were importing far less, putting downward pressure on world prices. Saudi Arabia, with the least costly oil to produce, kept pumping to keep prices low, hoping to put pressure on the more expensive North American oil producers. They had some success.

In the bust, producers became way more efficient and, like much of the US economy, found ways to operate with fewer people. Drillers could drill profitably with oil at $45. Production was steadily increasing. More oil was coming out of the ground now than at the height of the boom. While each well requires fewer roughnecks, more wells were coming on line.

That changed rapidly in 2020. With the cratering economy triggered by the coronavirus outbreak, demand for oil dropped. At the same time, following their 2014 playbook, Saudi Arabia instigated an oil price war, in which they added a vast amount of oil to the global supply. These two events combined to drive the benchmark West Texas Intermediate price to below $20 in March 2020 and even briefly negative the following month. Companies at work in the Bakken were quick to begin cutting their budgets for new wells and rigs. But Williston is well accustomed to the volatility of the oil market.[8]

Regardless, the oil will be gone sooner or later, and the industry will walk away. In the meantime, profits from the Bakken will largely go out of state, to oil companies and their stockholders, while costs are localized. The damage will remain. This drama has played out many times before. In Pennsylvania today, acid mine waste from coal mines closed a hundred years ago leads to red waterfalls and streams barren of fish. In North Dakota, land damaged by brine spills will take years to recover. It's not clear how long oil well casings and plugs will last, but what is certain is that there will be tens of thousands of wells left when the oil companies leave. George Mitchell's "damn cowboys" will have gotten their extra percent. In his account of the oil boom, Russell Gold observes:

The unfettered market is the ugly beauty of the US energy system. You can usually count on this free-market approach to deliver needed energy, but it can also make you want to turn away and not look too closely.[9]

One part of the cost of drilling isn't localized. The same technology that generated jobs in the Bakken and altered the global energy balance also exposed vast new reservoirs of carbon, sealed off for millions of years, to the atmosphere. The giant pipelines leading from the Bakken are pouring out hydrocarbons laid down in the shallows of that ancient ocean far back in geologic time. That carbon gets burned, goes into in the air, and spreads globally through the atmosphere, where it will stay for hundreds of years. It won't be going anywhere anytime soon, and it fuels the accelerating warming of the planet. The shimmering heat I felt riding into Williston will become well-known to future generations.

As I walked out of the Rib Fest, I noticed writing in multicolored kid's chalk on the sidewalk: SAVE THE EARTH.

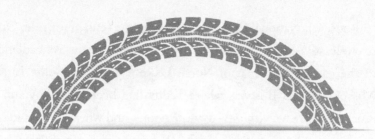

15

THE CORE OF THE CORE

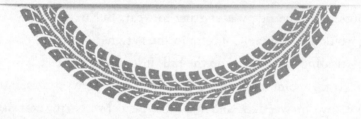

looked out the window at the back of Teddy Roosevelt's head.
I was at the Roosevelt Inn in Watford City, the next big town
south of Williston, bound for Theodore Roosevelt National
Park. In front of the motel was a white, 20-foot tall bust of the
former president. Over Teddy's hair, between his park and me,
were flames on the horizon. Oil wells flared off excess natural
gas, as on the approach to Williston. The website of Continental
Resources, the largest oil producer in the Bakken, refers to the area
I was observing, the day's route, as "the core of the core."

Roosevelt National Park seemed the natural end of my ride. One of our least-visited national parks, it's a stunning expanse of badlands in the southwest corner of North Dakota, watered by the Little Missouri River. This was where Teddy had been a cowboy and a rancher, where he withdrew to write books, and where his first ideas on conservation were formed. The Park Service runs a video about this remarkable place called "Refuge of the American Spirit." I was ready for that, though it was still two days down the road.

During one stretch while he was ranching, Teddy spent thirty-two days straight on horseback, covering over a thousand miles. At Watford City, I was approaching a thousand miles on my own horse, and the ride was starting to wear. But in just a few days, I would be meeting Concetta in the national park, and that was motivating me to pedal up the hills just a little quicker.

At this point in the ride, leg muscles and sinews were standing out, my skin was dark, and in most respects I was in my best shape. Yet the body and the bike had a dozen little problems. Muscles twitched when I stopped. Numbness in my feet had me clipping out of my pedals by midday, and the numbness didn't seem to go away completely at night. Would that worn strap on the handlebar bag last another 200 miles? Was that little chunk of rubber out of the tire a serious problem or something to ignore?

It had taken me a day to ride the 47 miles south to Watford City from Williston. Early on a sleepy Sunday morning, I slipped out of Williston onto the plains under an azure sky, happy for the light traffic. A return to US 85, the gateway to the Bakken, was the day's route. The road is legendarily dangerous, but it's been widened since the boom days of 2011. Then again, I was here in the summer, the easy season, without the excitement of water trucks and waste trucks rumbling down glazed asphalt during a North Dakota winter.

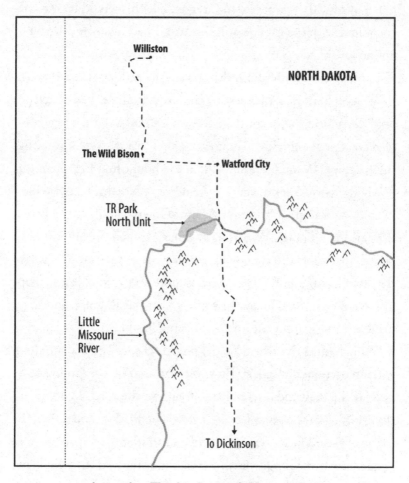

Approach to Theodore Roosevelt National Park.

On the outskirts of Williston, I rode across the main branch of the Missouri, the Big Muddy. In April of 1805, Lewis and Clark and the Corps of Discovery poled and rowed their way up the river past this place in their keelboat and pirogues, each stroke pushing them further into the unknown. I had followed their route for much of my 2011 ride west to Oregon, but this would be my only encounter with the captains on this trip. I had my own little bit of unknown ahead.

I pulled up a steady climb rising from the valley of the Missouri, Williston's Miracle Mile receding into the distance. I was back out onto the prairie, but in the distance was a strange sight: a graveyard of recreational vehicles.[1] "Recreational" is an odd term for what these campers were used for. When the boom first hit, there was nowhere for workers to stay, so spending a North Dakota winter in an RV was their only option. Many of them had likely camped in the parking lot of the Williston Walmart. When the oil bust happened, many left the area, walking away from the campers where they had lived. The RV graveyard is quite an eyesore, but perhaps it's also a sign of stability. In some cases, the RVs are no longer needed because more affordable housing has become available.

Visible amid the wheat fields, pumpjacks and tanks sprouted, with an occasional derrick. They were familiar on the landscape by now. If the drill pads were close enough to the road, I could hear the creaking of the pumpjack as it slowly nodded up and down, the music of the oil field. The pads were easy to spot from space as well. A quick look on Google Satellite that night showed the telltale red clay rectangles spread across this part of North Dakota.

I had almost resigned myself to a meal of roadside granola bars and Gatorade when I came upon the Wild Flour Cafe in Alexander. People were gathering for Sunday lunch after church.

Cowboys and roughnecks, often one and the same, sported Stetsons and baseball caps and faces lined from the sun. One leaned over the counter to banter with the cook. They began pulling together big tables with their families. Handshakes and stories were passed around, and a little girl in a high chair was exploring pancakes and syrup by hand. I missed home.

Not far down the road from Alexander is a dogleg bend on US 85 and a truck stop, the Wild Bison, where Bakken author Maya Rao worked. Men would shuffle in from twelve-hour shifts on the rigs, grab a pepperoni pizza and Mountain Dew, and watch TV in the lounge. It was, and probably still is, a rough corner. It's easy to imagine a truck driver riding miles down a straight road in the dark on little sleep and missing this turn. Multiple wrecks happened there, but traffic isn't the only excitement. Lightning struck an adjacent oil waste storage facility, and flames went 200 feet in the air. Amazingly, no one was hurt, and the Wild Bison survived intact.

Another 17 straight, uneventful miles brought me into Watford City. It doesn't have as much of the boomtown feel as Williston. Clearly the town has put some of the oil money into infrastructure. New stores populate the downtown, and all of those drill pads out on the plains paid for the new high school.[2] There's less the sense of men sleeping in cars and RVs. But on the way into town, fields full of mobile homes cover rolling hills. And those flames are a constant, tongues of fire always licking the horizon.

This is McKenzie County, the center of Continental Resources' Bakken operations. Continental is largely held by its founder, one larger-than-life man: Harold Hamm. Hamm was the youngest of thirteen children of Oklahoma sharecroppers. His first pair of new shoes did not come until he was five.[3] Uninhibited, energetic men populated the oil fields around his home town of Enid, and Hamm came to admire them. He started driving oil service trucks, then began his own operation. As a wildcat driller, he brought in his first well at age twenty-six. Lacking a formal degree, Hamm would come to know more geology than most geologists. His genial personality and down-home language disguised a curious mind and a driving ambition.

"I wanted to put myself in a position to find the ancient wealth. I had this burning desire," he said. "We all want to change the world and be bigger than our daily existence. I'm no different."[4]

Hamm and Continental would drill extensively in Oklahoma, but the biggest discoveries came in the Bakken play. An interesting word, "play." The vocabulary of the oil field is that of a great poker game. A driller lays down bets in terms of the vast expenditures for drilling. A modern fracked well can cost $18 million. Hold cards close to the vest: For the first commercially viable well in the North Dakota Bakken in 2004, Continental drilled under the name "Jolette Oil" so that competitors wouldn't know they were on to something. Then lay the hand down: By 2008, when the rest of the country was descending into the Great Recession, production was roaring in western North Dakota.[5] Though booms and busts would follow, oil development in the Bakken was off to the races, and Continental Resources was in the lead.

Hamm and Continental were persistent in the Bakken. They took risks when no one else would.[6] Those risks brought rewards. By 2018, Harold Hamm was one of the sixty wealthiest people in the United States, dubbed by *Forbes* a "shale legend."[7] Arguably, he's a case study of the American Dream.

Around the time that the Bakken was booming, the ground was beginning to shake back in Oklahoma, Continental Resources' other major play. In November of 2011, a 5.6 magnitude earthquake hit the small town of Prague, destroying sixteen houses. Until 2008, Oklahoma had experienced about two earthquakes over 3.0 magnitude per year. By 2014, 585 quakes would be rattling the state, triple the rate of California. They were routine enough that they were featured on weather reports and electronic billboards. Oklahoma was now the most seismically active state in the lower forty-eight. In 2016, another 5.6 magnitude quake would hit Cushing, known as the Pipeline Crossroads of the World.[8] Fifty-four million barrels of oil are stored there. It's to be the southern terminus to Keystone XL Pipeline.

What was going on? Katie Keranen, a seismologist at the University of Oklahoma, was lead author on a 2013 paper linking the Prague earthquake to the underground disposal of drilling waste from oil and gas operations.[9] The idea was that the large amount of waste fluid from Oklahoma's 3,200 disposal wells, injected near faults and basement rock, can cause them to slip. The response from the Oklahoma Geological Survey (OGS), housed at the University, was that the Prague quake was "the result of natural causes." Email archives from the university suggest that there was some questioning of OGS's objectivity. One researcher commented that the agency "couldn't track a bunny through fresh snow."[10]

But at least one member of the OGS staff was coming around to the idea that oil and gas activities might have something to do with Oklahoma's earthquake swarms. Later in 2013, seismologist Austin Holland put out a joint press release with the US Geological Survey that the chance of earthquakes of magnitude 5.5 or better had significantly increased. Shortly after the release, Holland received an email inviting him to have "coffee" with Harold Hamm and David Boren, a powerful former US senator and the president of the University of Oklahoma. After sharing the email with colleagues, one wrote back, "At least it's not Kool-Aid."

There were more than a few cross-ties. Hamm has donated more than $30 million to the University of Oklahoma, and his 2014 net worth was roughly double the Oklahoma state budget. Boren sits on the board of directors of Continental Resources. It's safe to say that this visit might be a little intimidating in the best of circumstances. Holland recalled that Hamm told him to "watch how you say things."[11]

Hamm would later claim he was never trying to influence the science. He said he was disturbed that an ill-advised report about fracking and earthquakes could be misinterpreted. "If you make a statement like that, it's going to cause a lot of concern," he said.[12]

I learned a lesson early in my career about suggestions from powerful people. I once drove ships for the National Oceanic and Atmospheric Administration. As a junior officer, I was piloting the ship on approach to a pier in Puerto Rico. We were coming in hot, and the captain, seated behind me, offered a thought.

"Don't you think we ought to back down a little, Dave?" he observed. I briefly rubbed my chin to contemplate the wisdom of his suggestion.

"Back down, god-dammit," came a rather more firm voice from behind.

Like my skipper, Harold Hamm became less subtle with time. Larry Grillot, Holland's immediate supervisor, documented a later meeting in which Hamm said "he would like to see select OGS staff dismissed."[13] Hamm also offered his services on the search committee for a new OGS director after Grillot's retirement. Hamm indicated that he would be speaking with Oklahoma Governor Mary Fallin about moving OGS out of the university. Austin Holland would later leave OGS. Katie Keranen left Oklahoma shortly after her paper was published.

Despite Continental Resources' best efforts, "induced earthquakes" became part of accepted science, with particular effort from the US Geological Survey. Oklahoma earthquakes peaked in 2015, declining in each subsequent year as waste disposal injections have been reduced. As of 2018, Oklahoma still has about seventy times more earthquakes than before 2008.[14] And six years after the earthquake, two oil companies settled with a Prague couple for an undisclosed amount for damage to their home.

One characteristic of a successful entrepreneur is single-mindedness. When presented with an obstacle, go by it, under it, or over it. In this case, the obstacle was a scientist. When confronted with the distinct possibility, later proven true, that Hamm's industry and his company were at least partly responsible for earthquake swarms in his home state, his response was to go after the scientist proposing the idea.

For Hamm, the obstacles extend to the national and international arenas. He delivered an endorsement speech for Donald Trump at the Republican National Convention. After President Trump's election, he was prominently mentioned as a candidate for Secretary of Energy. Hamm would presumably approve of much of what has happened since 2017. The United States withdrew from the Paris Climate Agreement. Federal environmental regulations, particularly as they apply to oil and gas, have been rolled back rapidly. President Trump immediately approved the Keystone and Dakota Access pipelines, which would provide a vast increase in pipeline capacity out of the Bakken as a side benefit.

There is a certain relentlessness, a drive to get every barrel of recoverable oil out of the ground. A *Bloomberg* interview described Hamm talking with his subordinates about the hunt for oil during the Bakken boom days of 2011:

> He wraps it up after signing off on a winter drilling budget in the Bakken. He sets his pen down and says, "I know y'all are turning over every leaf, but oil's up over a hundred." He grins. "What can I say?"[15]

I rolled past the dust and well flares outside of Watford City and rode through the core of the core. Somewhere around lunchtime and near the middle of nowhere, a large yellow drop outlined the sign for the Sweet Crude Travel Center. In oil terminology, light sweet crude is the most sought-after oil, as the greatest fraction of its volume can be processed into gasoline. I dodged the potholes in the parking lot.

Coming through the door was the smell of bacon and corn dogs under heat lamps. In the back were showers and a wall of shame for shoplifters caught on their video camera. A rack of evangelical books was at the head of an aisle of electronic chargers, baseball caps, and energy shots. A cornucopia of all possible varieties of jerky lay at my feet. I put a couple of burritos into the microwave and watched as cheese-bean goo oozed out onto the paper. I negotiated the puddle with plastic knife and fork, recharged my Gatorade, and took an uneasy stomach back out on the road.

After a long, flat ride through the oil field, the earth seemed to drop away. I rolled into a pullout, poised at the top of the canyon of the Little Missouri. A flat horizon was broken by great colored buttes, layered with yellow and brown sandstone, forests tucked into the couloirs. I was looking out at the badlands of the North Unit of Theodore Roosevelt National Park. In 1864, during a campaign against the Lakota, General Alfred Sully called the Little Missouri Badlands "hell with the fires out." Roosevelt called it the place where "the romance of my life began." Later, during his time in Washington, he would write to the western artist Frederic Remington, "I wish I were with you out among the sagebrush, the great brittle cottonwoods, and the sharply channeled barren buttes." Here I was.

A mile below in the distance, at the base of the hill, lay the sixty-year-old Long X Bridge. With a clear sightline and a break in the truck traffic, I peeled out onto my fastest descent of the ride: 40 miles per hour. Sometimes on a bike, speed is safety. If I could go fast enough on the winding descent, the trucks wouldn't try to pass me. I screamed across the bridge, trying to preserve my momentum for the big climb on the other side. As I flew across, I noticed that one of the overhead girders on the truss bridge was bent, almost snapped. It's not unusual. Oil supply trucks running over height have hit the Long X multiple times. A new bridge is in the planning stages. Then I bit into the big climb out of the canyon as the badlands opened up.

Halfway up the steepest climb of the ride, I was trying to keep my shoulder as close to the rock walls as I could on the winding road up the canyon. I needed to stay tight on the narrow road, but at slow climbing speeds the bike could get wobbly. My legs were screaming at me, but I needed to concentrate. I could hear the whine of the oil trucks coming up the grade behind me.

The phone rang in my front handlebar pack. I spotted a pullout and reached for the phone. It was Concetta, calling from the quiet of our house. Tomorrow she was flying into Bismarck, about 130 miles from where I was, to meet me. She's normally pretty calm, but I could hear frantic notes in her voice.

"There are no rental cars in Bismarck. The agent says there are no rental cars for 50 miles."

"Wait, how does this happen?" I asked. "Is there some kind of convention going on?"

"No, they say it was a monster hailstorm. All their rental cars were parked outside, and they've got a lot full of dents and cracked windshields."

"Don't worry. I'll find a car somewhere. There's no place you can go that I can't get to you." I would end up making a detour to the local airport at Dickinson and picking up a car there.

But the mention of the big hailstorm reminded me of a remark between a couple of farmers I'd overheard a few days before.

"Used to be we'd get these nice, gentle, all-day rains. Now it's like 'whoosh'—all in one big gully-washer."

In conservative North Dakota, I doubt they'd use the words "climate change." But people recognize that things are changing, here as everywhere else.

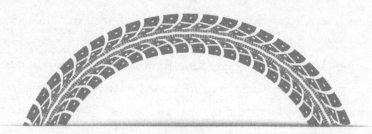

16

A CONVERSATION WITH TEDDY

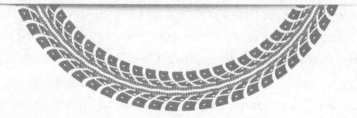

S omehow I always knew that Teddy Roosevelt would be waiting for me at the end of the road. When I was plotting out the ride back home, map windows sprawled across the screen, I noticed a green patch just south of the Bakken: his very own national park, one of the most remote and the only one named for a person. From the bottom of one of my mental closets, I remembered: This was the place where Teddy came to be a cowboy.

I didn't need to go to North Dakota to seek him out. One of Teddy's favorite places is one of mine as well: Rock Creek Park in Washington, DC. It's a little forested canyon, built up on all sides, meandering through the heart of the city. Grist mills lined the banks before the Industrial Revolution, and its cemetery was a hiding place on the Underground Railroad. For Roosevelt, it was a playground. He would drag guests on frenetic hikes through the park, leaving exhausted diplomats and cabinet members trailside. Roosevelt's rule for hiking, as in life, was this: You had to move forward "point to point," never circumventing any obstacle. Harold Hamm would approve. Roosevelt's dictum: "If a creek got in the way, you forded it. If there was a river, you swam it. If there was a rock, you scaled it, and if you came to a precipice you let yourself down over it."[1] In Rock Creek I could sense that this seemingly unlimited energy was nearby.

Before leaving for North Dakota, I took a training ride down the winding trail through Rock Creek and came upon one of Washington's most overlooked landmarks: the Jusserand Memorial. Jean Jules Jusserand, the refined, goateed French ambassador, was Teddy's hiking pal and erstwhile tennis partner, and one of the few who could keep up with him. Not quite overgrown and with a piece fallen off the front, Jusserand's marble bench sits in the woods along Rock Creek Parkway, without benefit of a parking lot. The monument marks the center of their rambles. Teddy's cousin Frank dedicated the bench in 1936. I pulled out some weeds growing from cracks in the marble.

Further down the trail, past where Rock Creek empties into the Potomac, is a forested island in the river. Like Rock Creek Park, it's an oasis in the middle of the city. Back in the woods, accessible only by bridge and footpath, stands the twice-life-size statue of Theodore Roosevelt, in the midst of a rousing speech, one arm raised to the sky. Vice President Roosevelt assumed the presidency after the assassination of William McKinley in 1901. Senator and political boss Mark Hanna famously fumed, "Now look, that damned cowboy is president of the United States!" How did Roosevelt earn that epithet?

Back in 1884, Theodore Roosevelt was the young, energetic, ascendant star of New York politics until an urgent telegram called him back on the train from Albany, the capital, to his New York City home. On Valentine's Day, on the same morning and in the same house, he lost his wife, Alice, to kidney disease and his mother, Mittie, to typhoid fever. By the end of the year, Roosevelt would leave his infant daughter in the care of his sister and board a train for North Dakota. The Badlands would be a place to mourn, to heal, to seek out solitude. He would find solace in the prairie that I'd ridden across, writing:

> Nowhere, not even at sea, does a man feel more lonely than when riding over the far-reaching, seemingly never-ending plains. Their very vastness and loneliness and their melancholy monotony have a strong fascination.[2]

Along the Little Missouri River in western North Dakota, Roosevelt came to practice what he called "the strenuous life." By the time he left two years later, he was not only healed but inspired, preparing him for the presidency and the time when he would alter the way that people thought about their wild lands. Roosevelt came west to pursue ranching, which he called "the pleasantest and healthiest and most exciting phase of American existence."[3] On the banks of the Little Missouri, he found a patch of land where he and his crew would build a house. Today at the National Park Visitors Center one can see his desk from that house, where he recorded his incredible adventures and experiences. He called this special place Elkhorn Ranch, arguably where the conservation movement was first imagined.

I finished the climb out of the North Unit of the Park from the Long X Bridge. Before long, I was cycling again through relatively flat prairie. A 67-mile day that began in Watford City ended in the relative civilization of Belfield, a town along Interstate 94. This was the last night of my bike trip. I'd been on the road for over three weeks, beginning with a mosquito-filled night after landing at Fort McMurray.

I was ready for the ride to be over. I could use a chance for the aches to go away, for the stretched tendons and bruises and scrapes and bites to heal. My skin held an occasional peel of sunburn here and there, and roadside burrs still buried themselves into hand-washed socks. But there's a road discipline, a sharp immediacy that you come to miss after time off the bike: the focus on the weather coming over the horizon, the beautiful afternoon thunderheads, the run for shelter. The learned background concentration on that patch of sand coming up ahead or that pickup coming from behind. The odd bond with the collection of tubes and spokes and rubber and steel, that old Trek that allowed me to stretch across a fair piece of the continent.

It had been very much a solo journey. I'd had to find my own way, solve my own problems, face my own fears. The only familiar face had been that of my friend Lynn Salvo, and that just for a day in Saskatchewan. All would change at the airport the next day. I would find my end of Teddy's never-ending plains. I missed company, Concetta's most of all.

There remained the task of picking up a car for Concetta and me. Dickinson Airport was 25 miles down the road the next morning and would be my last day on the bike. I started the morning like almost every other, stopping at a truck stop to pick up Gatorade. I imagined a sommelier behind the counter: *Would Monsieur prefer a bit of the red, or a bit of the green?*

This morning the cycling gods gave me one more little kick: a flat. I wanted nothing more than to knock out this last stretch to Dickinson, but a little something was sticking into my tire, and the road couldn't end until I found it. I pulled the tube out of the tire, re-inflated it, and passed it by my lips so I could feel the air coming out of the tiny pinhole. Eventually I found the piece of staple embedded in the tire. In the meantime, truckers walked by the concrete front pad, glancing bemusedly at the old guy in spandex swearing at the stripped-down tire, tube, and wheel. At least I wasn't wearing Teddy's tailored buckskins.

The airport lay a couple of hours down the road. It was small, modern, and nearly vacant. It doesn't often happen that I'm greeted as I walk in the door of an airport terminal. But I guess I was pretty recognizable, and a young man came out from behind the rental counter right away. They had the rental cars absent from Bismarck, and they were ready for me. After pocketing the car keys, I had to think for a moment and remember how to drive and how fast cars actually go. I disassembled the bike to put it into the back of the car. The Oxford English Dictionary defines the phrase "cowboy up" to mean "make a determined effort to overcome an obstacle." After 1,093 miles, I was ready to cowboy down.

I gradually got up to speed on the interstate for the hundred miles to the Bismarck airport. In an hour and a half I'd be covering what would take two days on the bike. A wind farm spun off in the distance.

Concetta emerged from security right on time and didn't mind a hug from me in my sweaty riding clothes. Later we drove back toward the expansive South Unit of Theodore Roosevelt National Park. We shared our first glimpse of the Badlands at an overlook off the interstate. The late evening light was beginning to bring out its colors. In a land carved by wind and water, pale pink layered buttes stretched to the horizon. A closer look at the hills revealed layers of yellow sandstone, black coal, and gray bentonite clay. At the foot of one of the buttes, a bison grazed.

We settled in for the night in Medora, the gateway town to the Park. Medora's carefully tailored western look includes the Rough Riders Hotel, where we lay our heads. At the front entrance, chandeliers hung from an inlaid ceiling. The concierge sat behind a wooden desk with a hand-carved front. On the desk at check-in was a sculpture of cowboys on horses, shooting guns into the air.

The next morning, for some reason having to do with three weeks on the road, I was ready to sleep in. Concetta, still on East Coast time, went out for a bird walk. On the way back, she made a discovery. Vegas has its Elvises. Medora has its Teddys.

Concetta had walked into Medora's sixth annual gathering of Theodore Roosevelt reprisers, with two Teddys, his son Quentin, and his aide Major Butt posing for a photo. Quentin was Teddy's youngest son, who dropped snowballs on diplomats from an upstairs White House window and was later lost in action over France during World War I. One of the other Teddys, Joe Wiegand, would later that afternoon perform a one-man show at the local theater. For every home state of every member of the audience, he had a story. He informed me from the stage that David Goodrich had been one of the Rough Riders. I imagine that Wiegand must have studied Roosevelt's patterns of speech from the herky-jerky early motion pictures that survive from that era on the Library of Congress website. He had all of the mannerisms: his neck thrust out, right fist smacked into left palm, then his head thrown back to laugh. He was "dee-*lighted!*"

Everything about Roosevelt seems larger than life, from the 20-foot bust in Watford City to Mount Rushmore to his island monument in the Potomac. A hundred years on, many of us tend to project our aspirations onto him. Senator Bernie Sanders, in announcing his tax proposal, quoted Roosevelt's warning about the dangers of a small class of enormously wealthy individuals "whose chief object is to hold and increase their power."[4] Alternatively, Senate Republicans invoked the former president in forming a new group concerning climate change and the environment. They called it the "Republican Roosevelt Conservation Caucus."[5]

Yet we worship him at our peril. In many ways Roosevelt was a ball of contradictions. His strongest legacy is in conservation, but he was a man who loved nothing more than killing large animals, lots of them. He won the Nobel Peace Prize but was the architect of the Spanish-American War. A racist and an imperialist, he invited Booker T. Washington to the White House during some of the worst times of the Jim Crow era. He said some things that are cringeworthy to modern ears, and to the ears of many of his own contemporaries.[6] For example:

> The most ultimately righteous of all wars is a war with savages, though it is apt to be also the most terrible and inhuman . . . it is of incalculable importance that America, Australia, and Siberia should pass out of the hands of their red, black, and yellow aboriginal owners, and become the heritage of the dominant world races.[7]

And yet he learned from the things he did. He originally came to the North Dakota Badlands to catch the last traces of the passing of the frontier age, when a hunter might still take the largest animal in North America. After a long hunt, Roosevelt succeeded in bagging a bison at a time when there weren't many left. After nearly killing one of the last ones, he became an advocate for their preservation. He sought to be a cattle baron from his Elkhorn Ranch. But the Badlands, like much of the West, were overgrazed, and his cattle were not fattened up for the disastrous winter of 1886. Roosevelt lost most of his herd to cold and starvation. That spelled the end for his ranch and a fair part of his fortune. He learned, and he brought those lessons east.

After becoming president, conservation became one of Roosevelt's main concerns. His disciple, Gifford Pinchot, became the first head of the newly created US Forest Service. He camped out for four days in Yosemite with John Muir, founder of the Sierra Club. Theodore Roosevelt protected approximately 230 million acres of public land during his presidency, including 150 national forests, five national parks, and eighteen national monuments.[8]

Concetta took me for a birding trip in his park. We shared a knife-edge ridge on the top of a bluff with a boy and his grandmother. He seemed to be trying to convince her of something. Just a few feet beyond us, Concetta spotted the unmistakable tube-shaped body, diamond patterns, and triangular head. We heard a soft buzz as it slipped into the brush. The boy was triumphant.

"I told you I saw a rattler! I told you!" he said.

We left our adventurers for a pebbly shoreline of the muddy Little Missouri. Four American kestrels sat in a cottonwood tree on the bank of the river. Swallows turned and zipped over the water against the backdrop of a great butte. We could imagine Teddy floating by this spot in the winter of 1886 on a makeshift raft.

Seeing photographs of Roosevelt during this era, the impression is of a strutting dandy from the East. In 1885, he posed with a rifle in a tailored buckskin suit. One might suspect that he wouldn't last a week out on a real ranch with real cowboys. In fact, it was the cowboys who would turn out to have a rough time keeping up.

In the late winter of 1885–6, Roosevelt was incensed to discover that his boat had been stolen from the riverside at the Elkhorn Ranch. In his fury, he resolved that the robbers wouldn't get away with it. In three days, he and his crew hammered together a flat-bottomed scow and launched it into the icy waters of the Little Missouri. They surprised the outlaws downstream, not far from our birding spot. With the thieves in custody, they proceeded down the river to a ranch. In camp in the evenings, Roosevelt was in the process of reading Tolstoy's *Anna Karenina*.

He arranged for a rancher to take his prisoners in a wagon to the sheriff. As their guard, he walked behind the wagon, in the cold, to the town of Dickinson, 45 miles away. After dropping off his prisoners, Roosevelt limped off to find the town doctor to bandage his blistered feet. The doctor described him as "all teeth and eyes. His clothes were in rags. He was scratched, bruised, and hungry, but gritty and determined as a bulldog."[9]

After our bird walk, I settled into a cozy chair among the book collection in the lobby of the Rough Rider Hotel, surrounded by Teddy memorabilia and enjoying the feeling of legs at rest. Later, Concetta and I stopped in for a drink on the patio, not realizing at first that it was the site of a private affair. Before long, a man approached our table. We were sure that we were about to get kicked out.

"Sorry to bother you," he said. "But our group is meeting here, and a couple of people didn't show up. So we have two spare tickets to dinner and the show at the Medora Musical. Would you like them?" We thanked him profusely.

We had dinner at the Pitchfork Fondue with our benefactor Kenny, a North Dakota water manager who was the great-grandson of a Dutch homesteader. The smell of steaks grilling on giant pitchforks drifted over the open-air dinner. Singers in the musical chatted with their future audience as they strolled between long picnic tables. We sat looking out over the colors of the Badlands at sunset while Kenny told us something about the agriculture side of North Dakota:

> Most small farmers are gone. In farming today, the machines are so large and expensive that small farms are less and less economical. Big operators are buying up the land and hiring crews to farm it. The small towns [besides those in the oil fields] are dying slowly. Where do all the kids go? Fargo, Bismarck, Minneapolis.

It sounded a lot like what I'd heard in Saskatchewan. The young people there were drawn to the lure of the small cities like Saskatoon and Regina. He continued:

> North Dakota [like Saskatchewan] has grown wetter over the last decades, but it's anything but consistent. Yet there's a steady trend toward putting more crops into production as opposed to rangeland. Hardly anything is irrigated in North Dakota.

Toward the end of dinner, I brought up climate change. Our discussion was brief.

"Climate's always changed and it always will," Kenny said. "We've had Ice Ages before, and I suspect we'll have them again."

"Okay, but I've spent some time on this," I replied. "We're well outside of natural variability and Ice Age cycles." Kenny started fidgeting with his fork.

"I suspect that we disagree on lots of things, but we can still talk."

"Yes we can," I said. And we moved on to the evening's entertainment.

The open-air Medora Musical is something of an institution, unfolding on a hillside amphitheater a few miles outside of town. The set is the archetype of an American small town, with dance numbers, cowboys on horseback, and trained elk on the hillside behind the stage. One of the set pieces, complete with pyrotechnics, is an on-stage recreation of Teddy's heroic charge up San Juan Hill during the Spanish-American War.

In the musical on the hill and in the town in the valley, Medora curates the myths, as do many of the rest of us. Off in the distance were the lights of the drilling rigs. I wondered how the real Teddy might have responded to the Harold Hamms of the world. If we were to have a conversation, what might he say?

Roosevelt had no knowledge of the oil below his ranch or of climate change. But the words of his presidency provide a hint of his thoughts in a speech that he titled "Conservation as a National Duty":

> We have become great in a material sense because of the lavish use of our resources, and we have just reason to be proud of our growth. But the time has come to inquire seriously what will happen when our forests are gone, when the coal, the iron, the oil, and the gas are exhausted, when the soils shall have been still further impoverished and washed into the streams . . . The time has come for a change.[10]

Roosevelt's reputation was not built just on conservation, but also on breaking up the great trusts that had come to rule the US economy after the end of the Civil War. From 1865 to 1900, when the Industrial Revolution was in full swing, power and wealth was increasingly concentrated in the hands of a few monopolies. The footprints of the leaders of these companies can still be seen in the row of "summer cottages" in Newport, Rhode Island: mansion after mansion, each larger than the last, standing as monuments to the period that came to be known as the Gilded Age. Dominating their respective sectors, the captains of industry were dubbed the robber barons: Andrew Carnegie for steel, J. P. Morgan for finance, and John D. Rockefeller for oil, among others.

Early in his first term, Roosevelt battled with J. P. Morgan in 1902 to break up his railroad trust, Northern Securities, then the second-largest company in the world. Morgan took the train down from New York to meet with President Roosevelt, as he'd done with presidents before. Morgan stated, "If we have done something wrong, send your man to my man and they can fix it up." Roosevelt replied, "That can't be done." Philander Chase Knox, his Attorney General added, "We don't want to fix it up. We want to stop it."[11] The Supreme Court broke up Northern Securities in 1904. Today the heights of the financial world are held, and the financing of the Keystone XL Pipeline provided, by the firm of JPMorgan Chase. It's also the top funder of fossil fuel extraction globally by a wide margin.[12]

A similar fate befell Rockefeller's Standard Oil, though it took the Supreme Court until 1911 to break it up. But the various pieces of the trust would be back. The Standard Oil of New York component, after mergers, became Mobil Oil in 1966. Standard Oil of New Jersey became Eastern States Standard Oil (Esso), renamed Exxon in 1972. Exxon and Mobil merged in 1999. The units were reaggregating. ExxonMobil is currently the fourth-largest producer in the oil sands.

On our last day in Medora, I looked up at the colored bluffs around town as I wheeled my bike into Dakota Cyclery, where they proceeded to box it up for the trip home. I'd gotten used to the long sightlines of the West, but I was ready to be back East. After a few days off the bike, the aches and pains had drifted away, and I'd even lost some weight.

I wasn't quite in Roosevelt's condition when he left to go back to New York in 1885. William Sewall, the Maine hunting guide who came to North Dakota to manage the Elkhorn Ranch, said simply: "When he got back into the world again he was as husky as almost any man I have ever seen who wasn't dependent on his arms for a livelihood. He weighed 150 pounds, and was clear muscle, bone, and grit."[13] When he returned to North Dakota in 1886, Roosevelt was invited to give the Fourth of July speech in Dickinson, the town where he'd brought the outlaws and I'd ridden to the airport. He was just warming up on a theme that he would speak about for a lifetime: "So it is particularly incumbent on us here today so to act throughout our lives as to leave our children a heritage, for which we will receive their blessing and not their curse."[14]

I suspect that Medora might look a little different the next time I come by. In a Park Service video, Theodore Roosevelt National Park is referred to as "an island in a sea of development." A new refinery is planned 3 miles from the park entrance, though court challenges and the oil price collapse has made acquiring the funding for the project difficult.[15] Up the Little Missouri from the park's South Unit, 180 new wells have been approved.[16] A new bridge, primarily for the benefit of those wells, is projected to service as many as a thousand trucks a day crossing the river. Seemingly without any sense of irony, NP Resources, the company doing the drilling, has dubbed the venture the Elkhorn Project.

Arguably, Elkhorn Ranch is where the American conservation movement began. Today, all that's left of the ranch are its foundations. They sit on a tiny outpost of the National Park, surrounded by Little Missouri National Grassland. Administered by the Forest Service, this is the largest grassland in the country, at over a million acres. That might imply some kind of protection. But as it says on the Forest Service signs, this is the "Land of Many Uses," and the high-value use of this land is drilling. Some 90 percent has been leased for oil and gas.[17]

Terry Tempest Williams wrote about Elkhorn Ranch in *The Hour of Land*, her work on the national parks. She took a ride up to the ranch site with Valerie Naylor, then the park superintendent. One of Naylor's most difficult jobs was to try to preserve the wilderness view that so inspired Roosevelt. She told Williams:

> I'm on border control every day of every week, trying to stop the rigs from going up in our view shed. I've completely given up on our governor and the legislature. They're very tight with the energy companies. So I've got some folks who alert me when there is a proposal for new development, and then I go directly to the CEOs

of the company and ask if they will meet with me. I've had my best luck working directly with the oil companies because our state regs are so poor. But I'll tell you, honestly, it's relentless and depressing and I'm tired.[18]

The First Amendment provides for the right of citizens to petition the government for redress. In North Dakota the government petitions the oil companies. Superintendent Naylor retired in 2014. For an update on her battle, I took a quick look on Google Maps and toggled over to satellite view. The telltale red patches of drill pads were easy to pick out. Seventeen are within 3 miles of Elkhorn Ranch. Even without the 180 new wells, there are drill pads on all sides.

Theodore Roosevelt died in 1919. A hundred years after his passing, Teddy is surrounded.

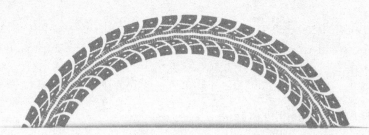

17

A CERTAIN RELENTLESSNESS

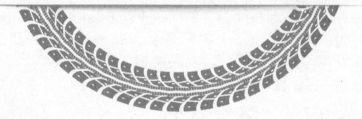

The oil and gas industry is too big to wrap your head around. The gas flaring that I saw in the Bakken has its counterparts in Nigeria, Algeria, and Siberia. The industry shapes global economies from Venezuela to Iran, from Russia to the United States. It's way bigger than one country, so I thought I'd look at two, and see how it manifests itself in two different nations. My route shadowed much of the Keystone XL pipeline, which has linked the oil and the politics of the United States and Canada. In the United States, drilling pads envelop the birthplace of conservation in Theodore Roosevelt National Park. After riding through this northern dominion of oil, I can understand the word that Valerie Naylor used to describe the approach of drilling to her park: relentless.

The View from the Ground

Every family has one: the person who will pose the direct, undiplomatic questions. When I was planning the trip, I talked with my sister-in-law. She's known for extraordinary (and sometimes useful) bluntness, like asking our son and his wife, "So when are you two thinking about having a baby?" We were chagrined at the directness of her question, but quite interested in the answer.

Thus, when I told her about my plans to ride from one oil field to the other, the unvarnished interrogation came out:

"So what do you want to do? Make them stop?"

"That would be good," I said.

It's true. It will come as no surprise that, as a climate scientist, I view the unlimited burning of fossil fuels as a slow-motion disaster. After all, I marched on the Mall in Washington, DC, and helped carry a pipeline in protest around the White House. Yet as the late journalist Cokie Roberts said, "Stories are simple until you cover them." I would add a corollary: Stories are different from ground level.

After doing long bicycle tours for almost twenty years, I continue to be delighted by the kindness of people along the way. In Fort McMurray, Bryan invited me into his pickup to tour the oil sands operation, something he clearly viewed with pride. Jesse opened her car during the tempest in the boreal forest. Doris at the Purple Palace offered me an oasis in the heart of the prairie. Down in the States, I had a chance to watch a large Tongan with a small baby at the Williston Rib Fest and realize that people with families are trying to build a life in the oil patch.

The boomtowns were fascinating. Fort McMurray and Williston are Canadian and US realizations of the same frontier oil economy, with a few differences. Fort McMurray has all of the pickups and none of the guns of Williston—that I could see, anyway. You'd be hard pressed to find rifle door handles on a bar in Canada. Williston has a little more of the frontier feel than Fort McMurray, but the rise of fracking is also more recent than the growth of the oil sands. Regardless, Sacramento and San Francisco started as boomtowns once upon a time during the gold rush. Mark Twain's journal of his gold rush days was called *Roughing It*. He would feel quite at home in Williston.

Bryan was quick to point out that Fort McMurray's oil patch supports not just the town but a fair piece of the Canadian economy. About 4 percent of Canada's total employment is in the energy sector.[1] In North Dakota, oil has also made a big difference. In 2008, a *National Geographic* photo essay on the state titled "The Emptied Prairie" featured images of abandoned farms.[2] Today, western North Dakota is anything but emptied. Though I saw my share of declining towns and tottering barns, oil money seems to be making a difference. In 2019, my friend Lynn Salvo cycled north through four states of the US Great Plains—Kansas, Nebraska, North Dakota, and South Dakota—and found the North Dakota towns distinctly better off than those of the other three.

The prosperity comes with a cost. Andy Skuce, a retired oil company executive, articulated the central tragedy of the climate change problem:

> The benefits of burning fossil fuels are concentrated on relatively few people and are fleeting, yet the environmental costs are globally dispersed and are spread over millennia. As long as the atmosphere is used as a cost-free dumping ground by producers and consumers of fossil fuels alike, then the harm from this practice is going to be passed on to others, especially the poor and the unborn.[3]

It's analogous to maxing out your credit card month after month, with the minimum payment due in twenty years. Except that there's no credit limit, and someone else will get the bill.

Other long-term factors have been in play. The goal of US energy independence, the distant dream of the 1970s, is nearing reality today. The deployment of hydraulic fracturing has led to the United States becoming the largest crude oil producer in the world. Setting aside the waste issues, fracking itself is a remarkable technical achievement. Rig crews are routinely steering drills into thin, undulating layers of oil shale several miles down.

Along the road, I had a number of conversations about climate, usually short. The immediate implication was that climate was a political issue, something not to spend too much time on in a conversation with a stranger. The "climate's always changed" phrase seemed ubiquitous, almost like a mantra. I'd heard it more than once. At the Teacher's Lounge bar near the Canadian border, the

father-and-son oil field mechanics brought out the phrase immediately after I mentioned that I was a retired climate scientist. I
heard it from the water manager in Medora. Sometimes it's a bit
more elaborate. On our ride through the oil sands, Bryan asked:
"Did you see the thing on Nature about black holes, that maybe
they're pulling on the sun, and maybe that's what's causing global
warming?" I had to admit that I hadn't heard that one, but it did
seem like one of the more creative efforts to deflect blame. More
typically, the "climate's always changed" conversation is brief, with
a touch of defensiveness.

The idea that the steady warming of the planet is the result of
natural cycles is pretty easy to debunk. One possible culprit is the sun,
which certainly varies in the amount of energy that it sends to the
Earth. Since the 1970s, when warming really accelerated, solar radia-
tion has actually decreased. Volcanoes are sometimes cited as a major
source of greenhouse gases. Yet the amount of carbon that they give
off is only about 1 percent of the carbon dioxide emitted by humans
and about 15 percent of the methane. Variations in the Earth's orbit
indeed caused the Ice Ages, but these variations happen over tens of
thousands of years. Their effect over the last century is quite small.[4]

For much of the public, and for many of the people in the oil
industry, everything about climate is uncertain. There's a reason
for that.

The Fog Machine

The evidence for climate change, and the human role in
it, only grows stronger with time. The latest report from the

Intergovernmental Panel on Climate Change shows accelerating sea level rise, primarily due to the increased melting of the Greenland and Antarctic ice sheets.[5] Global temperatures have been rising steadily on a background of year-to-year variability. More rain now comes in big, intense events, and increases in greenhouse gases have contributed to increases in Atlantic hurricane activity.[6]

Yet despite the perception of widespread scientific uncertainty, the basic science of human-induced climate change is simple, straightforward, and has been understood for more than thirty years. Numerous scientific panels had issued consensus reports by 1988. The case was dramatized in that year by James Hansen of NASA, the same atmospheric scientist who would later declare the development of oil sands and shale oil as "game over" for climate. That summer, in headline testimony before Congress, Hansen stated that the warming trend could be detected "with 99 percent confidence." He was bold at the time, but not out of line with the science. History has borne him out.

The next year, 1989, was my first in the NOAA climate office and a time of remarkable change. James Baker, George H. W. Bush's incoming secretary of state, used his first speech to say, "We can probably not afford to wait until all of the uncertainties about global climate change have been resolved. Time will not make the problem go away." National and international organizations were being set up to address the issue. The IPCC was inaugurated in 1989, and the office that I would later head, the US Global Change Program, was legally codified the following year. In the preceding decade, oil companies, including Exxon, had set up research programs and considered how to grapple with the problem.[7]

Working in a government science agency, I couldn't advocate policy actions. Our job was to produce the scientific facts, and just the facts. But in the back of our minds, we thought that if we highlighted the problem, the political system might work to address it. It wouldn't end up working that way.

Around the time of Hansen's 1988 testimony, something fundamentally changed the course of the narrative. Oil companies realized the potential effect of emissions restrictions on their profitability. The American Petroleum Institute, along with the US Chamber of Commerce and fourteen other trade organizations, formed the insidiously named Global Climate Coalition. It briefed politicians friendly to the industry and enlisted scientists skeptical about global warming to write paid op-eds. Despite the overwhelming scientific support for the idea of human-caused climate change, media coverage of the climate issue typically assumed the form of a debate, with pro- and anti-warming scientists on opposite sides. It all seemed very confusing. The fog machine was working. By 1997, the Senate voted 95–0 against the Kyoto Protocol, which would have committed developed countries to greenhouse gas reductions of 5 percent. This was the forerunner to the 2015 Paris Agreement, where 197 nations adopted voluntary reductions in their greenhouse gas emissions.

By 2002, oil industry lobbyists occupied influential positions in the federal government. That year the memo from political consultant Frank Luntz stating the need to "make the lack of scientific certainty a primary issue" was delivered to the White House. Phillip Cooney, the White House aide who inserted the "significant uncertainty" language in the climate annual report came from the American Petroleum Institute and later left for ExxonMobil.

The Global Climate Coalition was disbanded in 2002, its sponsors embarrassed by its tactics. Its successors are still around. The Competitive Enterprise Institute (CEI), whose supporters include the Charles Koch Institute and the American Fuel and Petrochemical Manufacturers, disputes that climate change is a problem.[8] The head of CEI's environment program, Myron Ebell, led the Trump Administration's transition team at the Environmental Protection Agency and spearheaded the opposition to the Paris Agreement.

After many decades of climate science and observations, denial of human-induced climate change has become virtually impossible to sustain. Another argument is often raised: guilt. Alberta author Chris Turner articulates this in the conclusion of his book *The Patch: The People, Pipelines, and Politics of the Oil Sands*:

> There's more to the debate, though, more to the way forward, than being right. A thing of such scope and power and wealth as the Patch* doesn't go away overnight or in a few years. Building the entire industrial basis of modern society on a new energy regime does not happen overnight or in a few years. We will have the Patch for decades. It is a fixture on the Canadian

---

* From Fort McMurray to the Bakken to the Permian Basin, the area where oil is extracted is known as the Oil Patch.

landscape. It employs tens of thousands. It is a truly national project, or else the term has no meaning. We have all benefited. We are all stakeholders. We are all complicit.[9]

I'll add once more the confession: I drive, I fly, and I heat my house courtesy of fossil fuels. By this logic, I am also complicit, along with everyone else who uses fossil fuels. And if everyone is to blame, then no one is to blame. If we are all cast as stakeholders in the global fossil fuel enterprise, then there's no particular reason for any of us to seek to slow it down.

In that sharing of guilt is the water popping out of manholes in Miami Beach, the relentless rain of our new hurricanes, and yes, the apocalypse of the Fort McMurray fire. Our acceptance of this collective guilt leads to the surrender to a new and very different planet, one, for starters, with a lot fewer coastal cities.

The familiar adage is, "If you find yourself in a hole, stop digging." If no one's to blame, that's a convenient reason to keep digging.

The Bubble in the Earth

I've watched men (and it's mostly men) driving massive dump trucks in the oil sand mines north of Fort McMurray, and I've seen men on drilling rigs in the Bakken. I've been one of those men; my rigs were in the Louisiana offshore thirty years ago. It's all part of a giant machine, on autopilot, bringing the steamy climate of the Cretaceous back to our doorstep.

Yet it's not quite autopilot. If you ask the biggest boosters of fossil fuels about climate change, the answer is universally to ignore or change the subject. When Harold Hamm is questioned on climate, he talks about California irrigation or Islamic terrorism.[10] There's the Total CEO's disavowal of "I am not in charge of the planet."[11] And the United States left a vacant chair at a discussion of climate at the 2019 G-7 meeting of leading industrial nations.[12] The most charitable view is that the explosion of fossil fuel exploration in North America is all about energy independence, about the United States in particular not relying on offshore oil, a goal that has eluded US presidents since the Arab oil embargo of 1973. Not coincidentally, it's also a generator of vast wealth concentrated in relatively few hands.

The public face of the oil companies is about environmental concern. After watching their television commercials, one might be forgiven for believing that their principal undertaking is growing algae and saving butterflies. For the oil companies, there has been a gradual, grudging, thirty-year acceptance that manmade climate change is real. It's been a profitable and damaging wait. Since Hansen's 1988 testimony, more carbon has been released into the atmosphere than in the entire history of civilization preceding it.[13] ExxonMobil today even publicly favors a carbon tax, though has expended few of its substantial resources to make that happen.

As in most human endeavors, it's wise to look at what people do, not what they say. Not surprisingly, despite a global climate agreement, major global oil and gas companies have recently approved vast new exploration projects throughout the world. The Carbon Tracker Initiative, a financial think tank focused on energy transition, analyzed these projects against the goal of the Paris climate accords: keeping global temperature increase below 2°C (3.6°F). Very few are economically viable in this context, including all oil sands projects. Much of the US fracking potential may also prove too expensive to exploit in a low-carbon world, according to the report.[14]

Either the big projects get slowed or stopped, or the planet's temperature will stay on its inexorable rise. As discussed earlier, roughly 80 percent of the carbon in known reserves has to stay in the ground for the global temperature increase to stay below 2°C. If that happens, we have stranded assets, a great bubble of carbon under our feet. Consider the behavior of markets in another celebrated bubble—housing. *Washington Post* economic reporter Matt O'Brien commented on the psychology prior to the Great Recession of 2008 that began with the collapse of the housing market:

> The idea being that the more people are worried about something, the more they should do to try to avoid it—right? You'd certainly think so, but not always. Consider the housing bubble: Economists including Paul Krugman and the Center for Economic and Policy Research's Dean Baker spent years warning about the impending danger, but it didn't matter. Policymakers didn't do anything, and everyone else was *too busy trying to get in while the getting was good* to concern themselves with whether it was sustainable. [emphasis mine][15]

What's driving the relentless search for more fossil fuels? "I need to get mine." It's not so different from the T-shirt I saw in

Williston: PLEASE LORD GIVE ME ONE MORE OIL BOOM. I PROMISE
NOT TO BLOW IT THIS TIME. For the housing bubble, investors
assumed that there would always be someone willing to pay a
higher price. This was the ominously named "greater fool theory."
At some point, the theory collapsed. Sooner or later there wasn't a
greater fool willing to buy, and the bubble burst. So it will be for
fossil fuels. Governments and/or markets will eventually intervene.

That market intervention is already beginning. In January
2020, the giant asset management firm BlackRock announced
that it was beginning to redirect its investments, explaining that
climate change has put the world "on the edge of a fundamental
reshaping of finance."[16] Because of climate risk, BlackRock stated
that "we will see changes in capital allocation more quickly than
we see changes to the climate itself." More specifically, along with
BlackRock, foreign investors including HSBC and insurance giant
The Hartford have announced a halt to investing in or insuring the
oil sands. Like the tire on my bike, fossil fuel finance is starting
to lose air.

In one critical aspect however, the fossil fuel bubble is different from housing. A decade after the housing bubble burst, the
bubble and the Great Recession were largely forgotten. Much of
the economy had been rebuilt. But barring massive carbon dioxide
extraction, using some combination of technologies—as yet unper-
fected or un-invented—the carbon being extracted by hundreds of
giant dump trucks and tens of thousands of oil rigs will remain in
our atmosphere for centuries.

There's a certain convenient ignorance—actually a deliberate
ignorance—among people who should know better. I think that

deep down, people in the oil and gas industry know that climate is changing in a fundamental way, and that the burning of fossil fuels is largely responsible for it. Peel back the flimsy, discounted arguments, and there's a word that emerges: naive. As in, one is naive to believe that we can really change this fossil fuel machine that's so deeply embedded in the metabolism of the global economy. The governments have been bought and paid for. We're destined to screw up the planet, and you're naive if you believe otherwise. And by the way, let me get mine before the whole thing falls apart.

It's a little late. Things are falling apart now, even on the doorsteps of the places that are generating the problem. In 2012, Hurricane Sandy blacked out lower Manhattan, the financial center for large parts of the fossil fuel enterprise. A firestorm burned out a large part of Fort McMurray in 2016. Houston, known as the Energy Capital of the World, was hit by 51 inches of rain during 2017's Hurricane Harvey and another 40 inches in 2019's Tropical Storm Imelda. In each case, one can credibly say that climate change didn't do the whole job, but it certainly loaded the dice. As the biblical phrase goes, having sown the wind, we are in the process of reaping the whirlwind. The impacts of climate change are happening in the heart of fossil fuel development, along with the rest of the planet.

I'd say these effects of climate change have long been predicted by the science community, but in fact there's a good case that we underestimated what climate change would do. Back in the '90s, the consensus reports saw climate changing slowly and ice sheets as stable features of the planet.[17] Today we see loss of Greenland and Antarctic ice at unimagined speed, fueling accelerating sea level rise. Perhaps we should have been paying more attention to some of the non–peer-reviewed statements, like Wallace Broecker's "The climate system is an angry beast and we're poking it with sticks."

Or James Hansen's assessment of the full utilization of unconventional (shale oil and oil sands) fuels: "Game over for climate."

Or the statement from Indigenous leader Mike Wiggins Jr.: "Seems like those people don't want to hang around another thousand years."

In different worlds, from vastly different perspectives, they're saying much the same thing.

For today, for what we do now, the key is whether we can actually change the way we generate energy—away from fossil fuels—fast enough to avoid the worst of climate change. Are we naive to believe that the entire energy system can change rapidly? Entrepreneur Tony Seba, in his presentation entitled "Clean Energy Disruption," shows two photographs at the start.[18] The first is of the Easter Parade of 1900 on Fifth Avenue in New York City. The audience is encouraged to find the automobile amid the throng of horses. The next photo is of the same street in 1913. This scavenger hunt is "find the horse" (there is one, somewhere amid the traffic). Energy transformations can happen quite rapidly. Seba argues that, with the rapidly declining cost of wind and solar power, we are on the brink of another transformation of this scale.

In the wake of the coronavirus outbreak and the oil price war, the situation appears dire for the fossil fuel industry. But as anyone in the oil patch will tell you, boom-and-bust is the nature of the business. When the global economy revives, low oil prices will likely drive up demand. Larger gas-driven cars and trucks will suddenly be more affordable to operate, and the cycle will reset.

Or perhaps not. It may be that, given the fires, floods, and other climate shocks of the last decade, a younger generation begins to turn away from a lifestyle based on burning fossil fuel. Perhaps a silver lining to the devastating coronavirus-driven recession is that it bursts the carbon bubble. The massive jolt to the global economy might, in fact, catalyze the clean energy transition, to bring us to a place where the smokestacks and tailpipes of the last century are replaced by the whir of turbines and the hum of cars. Such dramatic change is not unprecedented.

A Distant Echo

Sitting in the lobby of the Rough Riders Hotel in Medora, by the stone fireplace, below the shelves of books by and about Roosevelt, I had a chance to consider the drill pads surrounding the Elkhorn Ranch and the planned refinery outside of town. We've been here before. He was here before.

Theodore Roosevelt came west to partake of the frontier and discovered a land barren of the great animals that he had come to hunt. He went back east to found a legacy of conservation. He lived in a time when the robber barons sat astride the economy. He sent J. P. Morgan, red-faced, on the train back to New York after telling him, "We don't want to fix it up. We want to stop it."

It's ironic yet perhaps predictable that the same trusts that Roosevelt broke up in the early years of the last century have re-formed as principal players in the dominion of oil. JPMorgan Chase is the world's largest financier of fossil fuel extraction. ExxonMobil, grandchild of the Standard Oil breakup, is the largest US-based oil company.

There is a scene in the second *Terminator* movie that comes to mind. The robot that is Arnold Schwarzenegger's character has just smashed the evil robot pursuing his teenage charge, after delivering his famous line: "Hasta la vista, baby." A thousand little metal pieces lie on the factory floor. Then, as the kid watches wide-eyed, the metal pieces melt and begin to flow back together. From the liquid-metal puddle, the robot begins to re-form just a few feet away. With a sense of understated urgency, Arnold says, "We don't have much time."

The IPCC would agree with Arnold. In their 2018 report, they argue that, in order to keep the world below a 1.5 degree threshold and avoid some of the worst impacts of climate change, greenhouse gas emissions need to fall by half by 2030 and to net zero by 2050.[19] The alternatives are fairly stark. The good news is that humans won't go extinct. We're remarkably adaptable. The bad news is that Boston, Charleston, Norfolk, New Orleans, Bangkok, Dacca, Jakarta, and Alexandria (both Egypt and Virginia) are in trouble from sea level rise. There's no particular upper boundary to how hot it will get, and no reason to expect that warming will stop at the end of this century, when most climate model projections stop. The *average* temperature of the Cretaceous period was 97°F, some 40° warmer than today.

So the global fossil fuel industry must be slowed and ultimately halted. Despite the current oil bust, that's unlikely to happen on its own. Oil companies will do what they do unless and until governments stop them. It has always been so. Were the companies of Carnegie or Morgan or Rockefeller ever large enough for them? Yet these relentless forces have been beaten before. At the dawn of the last century, Theodore Roosevelt brought an overwhelming vitality to the national scene, a human energy that even the cowboys couldn't keep up with. With a lot of help, Roosevelt broke the power of the trusts. He faced at least as big a challenge at the start of the last century as we do today. We can do this again.

On September 20, 2019, 7.6 million of our children were in the streets around the world in the Global Climate Strike demonstrations. On a bright, hot afternoon, Concetta and I joined some of them on the lawn in front of the Capitol. We were packed together closely, with a sea of signs so deep that we could barely see the stage: I'VE SEEN SMARTER CABINETS AT IKEA; SUMMER IS COMING; and THE EARTH IS HOTTER THAN MY IMAGINARY BOYFRIEND.

I dream that from that crowd, another hero will rise, like Teddy after his outlaw chase, "All teeth and eyes . . . gritty and determined as a bulldog." She might already be here. Perhaps I'll get the chance to meet her.

ENDNOTES

PROLOGUE: COWBOYS AND INDIANS

1 "Oil Sands Facts and Statistics," Alberta Energy, https://www
.alberta.ca/about-oil-sands.aspx.

2 David Hasemyer, "Enbridge's Kalamazoo River Oil Spill Settlement
Greeted by a Flood of Criticism," *InsideClimate News*, February 10,
2017, https://insideclimatenews.org/news/09022017/kalamazoo
-river-oli-pipeline-spill-settlement-epa-enbridge-oil-sands.

3 Ben Adler, "The inside story of the campaign that killed Keystone
XL," *Vox*, November 7, 2015, https://www.vox.com/2015/11/7
/9684012/keystone-pipeline-won.

1: THE TAKEOFF ROLL

1 D. Gordon, A. Brandt, J. Bergerson, and J. Koomey, 2015. *Know
Your Oil: Creating a Global Oil-Climate Index* (Washington, DC:
Carnegie Endowment for International Peace, 2015), 36.

2   "Oil Sands 101: Recovery," Alberta Energy, Government of
     Alberta, accessed March 2020, https://www.alberta.ca/oil
     -sands-101.aspx.

3   Gordon et. al., *Know Your Oil*, 19.

4   Elizabeth McGowan, "NASA's Hansen explains decision to join
     Keystone pipeline protests," *InsideClimate News*, August 29, 2011,
     https://insideclimatenews.org/news/20110826/james-hansen
     -nasa-climate-change-scientist-keystone-xl-oil-sands-pipeline
     -protests-mckibben-white-house.

5   "The United States is now the largest global crude oil producer,"
     Today in Energy, US Energy Information Administration,
     September 12, 2018, https://www.eia.gov/todayinenergy/detail
     .php?id=37053.

6   Robert Krulwich, "A Mysterious Patch of Light Shows Up in the
     North Dakota Dark," *National Public Radio*, January 16, 2013,
     https://www.npr.org/sections/krulwich/2013/01/16/169511949/a
     -mysterious-patch-of-light-shows-up-in-the-north-dakota-dark.

7   Dave Mosher, "At one end of Trump's revived Keystone XL
     pipeline there is a scene you must see to believe," *Business Insider*,
     June 24, 2017, https://www.businessinsider.com
     /keystone-xl-canada-oil-sands-photos-2017-1.

8   Bill McKibben, "Recalculating the Climate Math," *The New
     Republic*, September 22, 2016, https://newrepublic.com
     /article/136987/recalculating-climate-math.

2: THE SKY OVER THE OIL SANDS

1   Panel at Oil Sands Discovery Centre, Fort McMurray, Alberta, 2018.

2   Chris Turner, *The Patch: The People, Pipelines, and Politics of the Oil
     Sands* (New York: Simon & Schuster, 2017), iv.

3   David Thruton, "'Smells like strong cat pee': Smartphone app
     tracks oilsands odour complaints," *CBC News*, June 14, 2018,
     https://www.cbc.ca/news/canada/edmonton/wood-buffalo-air
     -quality-app-1.4703417.

4   Andrew David Thaler, "Alberta, Canada is the proud owner of the largest man-made pyramid on the planet," *Southern Fried Science* (blog), October 16, 2012, http://www.southernfriedscience.com /alberta-canada-is-the-proud-owner-of-the-largest-man-made -pyramid-on-the-planet/.

5   Turner, *The Patch*, 174.

6   Gabriel Friedman, "Oil plummets below break-even point for most Canadian producers—with more pain to come," *Calgary Herald*, March 18, 2020, https://calgaryherald.com/commodities /energy/oil-plummets-below-break-even-point-for-most-canadian -producers-with-more-pain-to-come/wcm/cd6c3364-7b75-4200 -8441-cc5be0f1e798/.

7   Canadian Association of Petroleum Producers, 2020. Oil Sands Reclamation," accessed March 25, 2020. https://www.capp.ca /explore/land-reclamation/.

8   Ed Struzik, "On Ravaged Tar Sands Land, Big Challenges for Reclamation" *Yale Environment 360*, Yale School of Forestry and Environmental Studies, March 27, 2014, https://e360.yale .edu/features/on_ravaged_tar_sands_lands_big_challenges_for _reclamation.

9   Pratish Narayanan, "Goldman Sachs Says U.S. Oil Drama Trumps OPEC's Vienna Chaos," *Bloomberg News*, June 25, 2018, https://www.bloomberg.com/news/articles/2018-06-25/goldman -sachs-says-american-oil-drama-trumps-opec-s-vienna-chaos.

3: APOCALYPSE THEN: THE 2016 FORT McMURRAY FIRE

1   "Wildfires," Get Prepared, Government of Canada, accessed 11/14/2018, https://www.getprepared.gc.ca/cnt/hzd/wldfrs -en.aspx.

2   David Staples, "Firestorm: How a wisp of smoke grew into a raging inferno," *Edmonton Journal*, May 1, 2017, https://edmonton journal.com/feature/firestorm-how-a-wisp-of-smoke-grew-into-a -raging-inferno.

3   KPMG Consultants, "May 2016 Wood Buffalo Wildfire Post-Incident Assessment Report," prepared for Alberta Emergency Management Agency: 25, May 2017.

4   Staples, "Firestorm."

5   KPMG, "May 2016 Wood Buffalo Wildfire."

6   Bill Kaufmann, "Outrage following Fort McMurray 'karma' tweet won't lead to job loss," *Calgary Herald*, May 6, 2016, https://calgaryherald.com/news/local-news/outrage-following-fort-mcmurray-karma-tweet-wont-lead-to-job-loss.

7   Dino Grandioni, "The Energy 202: Top House Democrats don't agree on who should lead on climate change," *The Washington Post*, November 15, 2018. https://www.washingtonpost.com/news/powerpost/paloma/the-energy-202/2018/11/15/the-energy-202-top-house-democrats-don-t-agree-on-who-should-lead-on-climate-change/5becc9f91b326b392905483c/?utm_term=.49e078958cfb.

8   KPMG, "May 2016 Wood Buffalo Wildfire," 18.

9   D. R. Reidmiller, C. W. Avery, D. R. Easterling, K. E. Kunkel, K. L. M. Lewis, T. K. Maycock, and B. C. Stewart, eds., *Impacts, Risks, and Adaptation in the United States: Fourth National Climate Assessment, Volume II: Report-in-Brief* (Washington, DC: US Global Change Research Program, 2018), 2.

10   "Climate Change and Fire," Natural Resources Canada, Government of Canada, accessed December 2018, https://www.nrcan.gc.ca/forests/fire-insects-disturbances/fire/13155.

11   Myles Allen et al., *Global Warming of 1.5˚C.: Summary for Policymakers* (Geneva: World Meteorological Organization, 2018), 6.

12   Reidmiller et al., *Impacts, Risks, and Adaptation*, 72–144.

13   McKibben, "Recalculating the Climate Math."

14   C. McGlade and P. Ekins, "The geographical distribution of fossil fuels unused when limiting global warming to 2˚C," *Nature* 17, (January 8, 2015) 187-190.

15 Turner, *The Patch*, 260–262.
16 Hannah Ritchie and Max Roser, "CO_2 and Greenhouse Gas Emissions," Published online at OurWorldInData.org. Retrieved from: https://ourworldindata.org/co2-and-other-greenhouse-gas-emissions.

4: VISIT TO A MAN CAMP

1 US Forest Service, *The Rising Cost of Wildfire Operations* (Washington, DC: US Department of Agriculture, 2015).
2 Scot Bransford, Jennifer Medina, and Jose A. Del Real, "As Carr Fire Kills 2 in California, Firefighters Reflect on a Job Now 'Twice as Violent,'" *New York Times*, July 27, 2018, https://www.nytimes.com/2018/07/27/us/california-wildfire-redding-carr.html.
3 Martin Kaste, "Can U.S. Take the Heat of Canada's Oil Practices?" *National Public Radio*, August 18, 2010, https://www.npr.org/templates/story/story.php?storyId=129277985.

5: "SIGNIFICANT UNCERTAINTIES"

1 Jennifer Lee, "A Call for Softer, Greener Language," *New York Times*, March 2, 2003, https://www.nytimes.com/2003/03/02/us/a-call-for-softer-greener-language.html. Text of Luntz memo can be found at https://www.sourcewatch.org/images/4/45/LuntzResearch.Memo.pdf.
2 "Rick Piltz—George W. Bush White House Climate Science Whistleblower," video by Government Accountability Project, November 19, 2013, YouTube video, https://www.youtube.com/watch?time_continue=170&v=64IuKM17n0w.
3 Andrew C. Revkin, "A Passing: Rick Piltz, a Bush-Era Whistleblower," *New York Times*, October 19, 2014, https://dotearth.blogs.nytimes.com/2014/10/19/a-passing-rick-piltz-a-bush-era-whistleblower/?searchResultPosition=6.
4 A. Leiserowitz, E. Maibach, S. Rosenthal, J. Kotcher, P. Bergquist, M. Ballew, M. Goldberg, and A. Gustafson, *Climate*

Change in the American Mind: April 2019 (New Haven, CT: Yale Program on Climate Change Communication, 2019), https://doi .org/10.17605/OSF.IO/CJ2NS.

5   William J. Ripple, Christopher Wolf, Thomas M. Newsome, Phoebe Barnard, and William R. Moomaw, "World Scientists' Warning of a Climate Emergency," *BioScience* 70, no. 1 (January 2020): 8–12, https://doi.org/10.1093/biosci/biz088.

6   Hiroko Tabuchi, "A Trump Insider Embeds Climate Denial in Scientific Research," *New York Times*, March 2, 2020, https:// www.nytimes.com/2020/03/02/climate/goks-uncertainty -language-interior.html?referringSource=articleShare.

7   Kate Yoder, "Top Republican message-maker Frank Luntz calls for climate action," *Bulletin of the Atomic Scientists,* July 26, 2019, https://thebulletin.org/2019/07/top-republican-message-maker -frank-luntz-calls-for-climate-action/#.

8   Nathaniel Rich, "Losing Earth: The Decade We Almost Stopped Climate Change," *The New York Times Magazine*, August 1, 2018, https://www.nytimes.com/interactive/2018/08/01/magazine /climate-change-losing-earth.html#epilogue.

7: THE WATER PROTECTORS

1   Amy Goodman, "First Nations Activist: Canada's Purchase of Kinder Morgan Pipeline Will Cause Environmental & Economic Disaster," *Democracy Now*, May 31, 2018, https://www .democracynow.org/2018/5/31/first_nations_activist_canadas _purchase_of.

2   Sarah Rieger, "Indigenous-led group submits unsolicited bid to buy Trans Mountain pipeline," *CBC News*, July 24, 2019, https://www.cbc.ca/news/canada/calgary/project-reconciliation -trans-mountain-1.5224374.

3   Kyle Bakx and Geneviève Normand, "Drawing a line in the oilsands," *CBC News*, December 25, 2018, https://newsinteractives .cbc.ca/longform/drawing-a-line-in-the-oilsands-fight.

4   Eriel Deranger, "I feel betrayed by the government and a system that has destroyed the spirit of my people," *National Observer,* April 24, 2018, https://www.nationalobserver.com/2018/04/24/opinion/i-feel -betrayed-government-and-system-has-destroyed-spirit-my-people.

5   Carol Christian, "Fort McKay First Nation: A Success Story," *Your McMurray Magazine* 6–4, May 22, 2018, https:// yourmcmurraymagazine.com/archives/features/2324/ fort-mckay-first-nation-a-success-story.

6   Shawn McCarthy, "Where oil and water mix," *The Globe and Mail,* November 6, 2015, updated November 12, 2017, https://www .theglobeandmail.com/news/alberta/where-oil-and-water-mix-oil -sands-development-leaves-fort-mckays-indigenous-community torn/article27151333/.

7   Eriel Deranger, "Reclaiming Our Indigeneity and Our Place in Modern Society," presentation to 2015 Bioneers Annual Conference, video by Bioneers, November 15, 2015, YouTube video, https://www.youtube.com/watch?v=KI89FHqvrf0 =.

8   Kevin Orland, "Teck Resources pulls Frontier oilsands mine project, blaming divisive debate over climate change," *Financial Post,* February 24, 2020, https://business.financialpost.com /commodities/energy/teck-pulls-application-for-oil-sands-mine -in-relief-for-trudeau.

9   Nick Estes, *Our History is the Future: Standing Rock versus the Dakota Access Pipeline, and the Long Tradition of Indigenous Resistance* (New York: Verso Books, 2019).

10 Nick Estes, Keynote talk for the 7th Annual Native American Alumni Dinner, April 7, 2017, https://nickestesblog, August 20, 2017 entry.

11 Estes, *Our History is the Future,* 10.

12 Estes, *Our History is the Future,* 11, 137.

13 Amy Dalrymple, "Pipeline route plan first called for crossing north of Bismarck," *The Bismarck Tribune,* August 18, 2016, https://bismarcktribune.com/news/state-and-regional/pipeline

-route-plan-first-called-for-crossing-north-of-bismarck
/article_64d053e4-8a1a-5198-a1dd-498d386c933c.html.

14 Estes, *Our History is the Future*, 48.

15 Louise Erdrich, "Holy Rage: Lessons from Standing Rock," *The New Yorker*, December 22, 2016, https://www.newyorker.com /news/news-desk/holy-rage-lessons-from-standing-rock.

16 Gregory Meyer, "Controversial Dakota Access pipeline boosts oil production," *Financial Times*, June 21, 2019, https://www.ft.com /content/29a60a66-92c4-11e9-b7ea-60e35ef678d2.

17 Emily S. Rueb and Niraj Chokshi, 2019, "Keystone Pipeline Leaks 383,000 Gallons of Oil in North Dakota," *New York Times*, October 31, 2019, https://www.nytimes.com/2019/10/31/us /keystone-pipeline-leak.html.

18 Winona LaDuke, "Prophecy of the Seventh Fire: Choosing the Path that is Green," lecture, Thirty-Seventh Annual E. F. Schumacher Lectures, Great Barrington, MA, November 2017, https://centerforneweconomics.org/publications/prophecy-of-the -seventh-fire-choosing-the-path-that-is-green/.

19 M.Collins, R. Knutti, J. Arblaster, J. L. Dufresne, T. Fichefet, P. Friedlingstein, X. Gao, W. J. Gutowski, T. Johns, G. Krinner, M. Shongwe, C. Tebaldi, A.J. Weaver, and M. Wehner, "Long-term Climate Change: Projections, Commitments and Irreversibility," in *Climate Change 2013: The Physical Science Basis*, ed. T. F. Stocker, D. Qin, G. K. Plattner, M. Tignor, S. K. Allen, J. Boschung, A. Nauels, Y. Xia, V. Bex, and P. M. Midgley (Cambridge, United Kingdom and New York: Cambridge University Press, 2013), 1103.

8: STEWARDS OF THE PRAIRIE

1 Art Cullen, "In My Iowa Town, We Need Immigrants," *New York Times*, July 30, 2018, https://nyti.ms/2NSSxQC.

9: THE BOMB ON THE RIDGELINE

1 "Oil in Alberta; Fuelish behaviour," *The Economist*, December 15, 2018.

2   Tim Murphy, ed., *Journey to the Tar Sands* (Toronto: James Lorimer and Co., 2008).

3   Meyer, "Controversial Dakota Access pipeline boosts oil production."

4   Kevin Orland, Historic price crash plunges Canadian oil patch into crisis," *Bloomberg*, November 20, 2018, https://www .bloomberg.com/news/articles/2018-11-20/canadian-oil-patch -plunged-into-crisis-by-historic-price-crash.

5   Ian Austen, "A Runaway Train Explosion Killed 47, but Deadly Cargo Still Rides the Rails," *New York Times*, July 16, 2019, https://www.nytimes.com/2019/07/16/world/canada/lac -megantic-quebec-train-explosion.html?searchResultPosition=1.

6   Gary McWilliams, "Exxon agrees to $1 million fine over 2011 Yellowstone River oil spill," *Reuters*, June 5, 2019, https://www .reuters.com/article/us-exxon-mobil-pipeline-montana/exxon -agrees-to-1-million-fine-over-2011-yellowstone-river-oil-spill -idUSKCN1T61VA.

7   James Hansen, "Game Over for the Climate," *New York Times*, May 9, 2012, https://www.nytimes.com/2012/05/10/opinion /game-over-for-the-climate.html?searchResultPosition=2.

8   Allen et al., *Global Warming of 1.5°C.*, SPM-4.

9   Global fossil fuel emissions of 34.9 billion tonnes of carbon dioxide in 2012 versus 37.2 in 2018: Corinne Le Quéré et al, 2018. "Global Carbon Budget 2018," *Earth System Science Data* 10, (2018): 1–54, https://doi.org/10.5194/essd-10-2141-2018.

10 *Banking on Climate Change: Fossil Fuel Finance Report Card 2019* (Oil Change International, 2019), 94, http://priceofoil.org/content /uploads/2019/03/Banking-on-Climate-Change-2019-final.pdf.

11 Jeffrey Jones, "Koch Industries sells its oil-sands properties to Paramount," *The Globe and Mail*, August 14, 2019, https://www .theglobeandmail.com/business/industry-news/energy-and -resources/article-koch-industries-sells-its-oil-sands-properties -to-paramount/.

12 "Statement by the President on the Keystone XL Pipeline," Office of the Press Secretary, November 5, 2015, press release, https:// obamawhitehouse.archives.gov/the-press-office/2015/11/06 /statement-president-keystone-xl-pipeline.

13 Peter Erickson and Michael Lazarus, "New oil investments boost carbon lock-in," *Nature* 526, no. 43 (October 1, 2105): https:// www.nature.com/articles/526043c.

14 David Roberts, "What critics of the Keystone campaign misunderstand about climate activism," *Vox*, November 8, 2015, https://www.vox.com/2015/11/8/9690654/keystone-climate -activism.

15 E. Bush and D. S. Lemmen, eds., *Canada's Changing Climate Report* (Ottawa, Ontario: Government of Canada, 2019), 1.

16 Tyler Dawson, "Alberta to investigate anti-oilsands funding." *National Post*, July 5, 2019.

17 Edmonton Journal Editorial Board, "We are voting for a stronger economy," *Edmonton Journal*, April 13, 2019, https://edmonton journal.com/opinion/editorials/editorial-we-are-voting-for-a -stronger-economy.

18 Amanda Coletta, "Canada's wildfire season is off to a ferocious start," *Washington Post*, June 11, 2019, https://www.washington post.com/world/the_americas/canadas-wildfire-season-is-off-to -a-ferocious-start/2019/06/10/9d2d67bc-88a2-11e9-9d73 -e2ba6bbf1b9b_story.html.

19 Jeffrey Jones, "'Our pipeline,' Notley's elusive Holy Grail," *The Globe and Mail*, 31 May, 2017, B1.

20 Kevin Taft, *Oil's Deep State: How the petroleum industry undermines democracy and stops action on global warming—in Alberta, and in Ottawa* (Toronto, Canada: James Lorimer & Company Ltd., 2017), Kindle edition.

21 Emma Graney, "Leader profile: UCP's Jason Kenney uses strong work ethic to find success," *Edmonton Journal*, April 15, 2019,

https://edmontonjournal.com/news/politics/leader-profile
-ucps-jason-kenney-uses-strong-work-ethic-to-find-success.

10: THE BIRDS OF SASKATCHEWAN

1   Susan Kieffer, "Rufa Red Knot Gets Listed," *Audubon*, December 11, 2014, https://www.audubon.org/news/rufa-red-knot-gets -listed.

2   Deborah Cramer, "Red Knots Are Battling Climate Change— On Both Ends of the Earth," *Audubon*, May–June 2016, https:// www.audubon.org/magazine/may-june-2016/red-knots-are -battling-climate-change-both-ends.

3   Alan R. Smith, C. Stuart Houston, and J. Frank Roy, *Birds of Saskatchewan* (Regina: Nature Saskatchewan, 2019), Special Publication No. 38.

4   Kenneth V. Rosenberg et al., "Decline of the North American avifauna," *Science* 366, no. 6461 (October 4, 2019): 120–124, https://doi.org/10.1126/science.aaw1313.

5   "Kitchener teen calls on Canada to declare national climate emergency," *CBC News*, May 21, 2019, https://www.cbc.ca/news /canada/kitchener-waterloo/kitchener-teen-calls-on-canada-to -declare-national-climate-emergency-1.5142339.

6   WHSRN Executive Office, "Red Knot population in Tierra del Fuego crashes to a new low," Western Hemisphere Shorebird Network, February 26, 2018, https://whsrn.org /red-knot-population-in-tierra-del-fuego-crashes-to-a-new-low/.

7   Sandra Díaz, Josef Settele, and Eduardo Brondízio, et al., "Summary for policymakers of the global assessment report on biodiversity and ecosystem services of the Intergovernmental Science-Policy Platform on Biodiversity and Ecosystem Services," IPBES Secretariat, May 6, 2019, https://www.ipbes .net/sites/default/files/downloads/spm_unedited_advance_for _posting_htn.pdf.

11: WHERE THE SEA USED TO BE

1   Edwin Dobb, "The New Oil Landscape," *National Geographic*, March 2013, 1–34, http://archive.nationalgeographic.com /?iid=76032#folio=CV1.

2   Corey J. A. Bradshaw and Ian G. Warkentin, "Global estimates of boreal forest carbon stocks and flux," *Global and Planetary Change* 128 (May 2015): 24-30, https://doi.org/10.1016/j.gloplacha .2015.02.004. Blog-post summary at https://ourworld.unu.edu /en/earths-second-lung-has-emphysema.

3   Peter McKenzie-Brown, *The Great Oil Age* (Calgary, Alberta: Petroleum History Society, 1993), 71.

4   B. Sames, M. Wagreich, J. E. Wendler, B. U. Haq, C. P. Conrad, M. C. Melinte-Dobrinescu, X. Hu, I. Wendler, E. Wolfgring, I. Ö. Yilmaz, and S. O. Zorina, 2015. "Review: Short-term sea-level changes in a greenhouse world—A view from the Cretaceous," *Palaeogeography, Palaeoclimatology, Palaeoecology* 441, part 3 (January 1, 2016): 393–411, http://dx.doi.org/10.1016 /j.palaeo.2015.10.045.

5   Mary Bagley, "Cretaceous Period: Animals, Plants & Extinction Event," *Live Science* (blog), January 8, 2016, https://www .livescience.com/29231-cretaceous-period.html.

6   Alanna Mitchell, "Why this stunning dinosaur fossil discovery has scientists stomping mad," *Macleans*, May 9, 2019, https:// www.macleans.ca/society/science/why-this-stunning-dinosaur -fossil-discovery-has-scientists-stomping-mad/.

7   Douglas Preston, "The Day the Dinosaurs Died," *The New Yorker*, April 8, 2019, https://www.newyorker.com/magazine/2019/04/08 /the-day-the-dinosaurs-died.

8   William Redekop, *Lake Agassiz: The Rise and Demise of the World's Greatest Lake* (Winnpeg, Manitoba: Heartland Associates, 2017).

9   Wallace S. Broecker, 1975. "Climatic Change: Are We on the Brink of a Pronounced Global Warming?" *Science* 189, no. 4201

(August 8, 1975): 460–463, https://science.sciencemag.org
/content/189/4201/460.

10   Wallace S. Broecker, "When climate change predictions are
right for the wrong reasons" *Climatic Change* 142, no. 1–6 (2017):
https://doi.org/10.1007/s10584-017-1927-y.

11   Richard Alley, "Abrupt Climate Changes: Oceans, Ice, and Us"
(lecture, Roger Revelle Commemorative Lecture Series, National
Research Council, Washington, DC, 2004), http://nas-sites.org
/revellelecture/past-lecturers/2004-2/.

12   Wallace S. Broecker, "The biggest chill," *Natural History* 96, no.
10 (1987): 74–82.

13   Wallace Broecker, James Kennett, Benjamin Flower, James Teller,
Sue Trumbore, George Bonani, and Willi Wolfli, 1989. "Routing
of meltwater from the Laurentide Ice Sheet during the Younger
Dryas cold episode," *Nature* 341, no. 318–321 (September 28,
1989): https://www.nature.com/articles/341318a0.

14   National Research Council, *Abrupt Impacts of Climate Change:
Anticipating Surprises* (Washington, DC: National Academies
Press, 2013), 1.

15   Stefan Rahmstorf, Jason E. Box, Georg Feulner, Michael
E. Mann, Alexander Robinson, Scott Rutherford, and Erik
J. Schaffernicht, "Exceptional twentieth-century slowdown
in Atlantic Ocean over turning circulation." *Nature Climate
Change* 5, no. 475–480 (March 23, 2015): doi.org/10.1038
/nclimate2554.

12: MEET ME IN MOOSE JAW

1   Reis Thebault and Emily Rauhala, "Intercontinental conflict
ends peacefully as Norway agrees Canada's got the bigger moose,"
Washington Post, March 6, 2019, https://www.washingtonpost
.com/world/2019/03/07/inter-continental-conflict-ends
-peacefully-norway-agrees-canadas-got-bigger-moose/?utm
_term=.30a4d8205730.

13: THE CROSSING

1 Average May–September (growing season) rainfall at Weyburn downloaded from Canadian Climate Centre January 2019.

2 Kathleen Norris, *Dakota: A Spiritual Geography* (New York: Houghton Mifflin, 1993), 12.

3 Maya Rao, *The Great American Outpost* (New York: Hachette Book Group, 2018), 11.

4 Rao, 286.

5 Likhitha Butchireddygari, "Salting the earth: North Dakota farmers struggle with a toxic byproduct of the oil boom," *NBC News,* August 11, 2018, https://www.nbcnews.com/news/us-news /salting-earth-north-dakota-farmers-struggle-toxic-byproduct-oil -boom-n895771.

6 Renée Jean, "Blacktail Creek spill anniversary comes and goes with yet another spill," *Williston Herald,* January 12, 2016, https:// www.willistonherald.com/news/blacktail-creek-spill-anniversary -comes-and-goes-with-yet-another/article_4722623c-b93e-11e5 -b365-abac3361137f.html.

7 Lauren Donovan, "Radioactive dump site found in remote North Dakota town," *Bismarck Tribune,* March 11, 2014, https://bismarcktribune.com /bakken/radioactive-dump-site-found-in-remote-north-dakota-town /article_39d0d08a-a948-11e3-8a3b-001a4bcf887a.html.

14: WILLISTON: BOOMTOWN, USA

1 Lawrence Wright, "The Dark Bounty of Oil," *New Yorker*, January 1, 2018, https://www.newyorker.com/magazine/2018/01/01/the -dark-bounty-of-texas-oil.

2 Dobb, "The New Oil Landscape."

3 J. D. Borthwick, *Three Years in California* (London: William Blackwood and Sons, 1857), 67.

4 Rao, *Great American Outpost*, 93.

5 Blaire Briody, *The New Wild West: Black Gold, Fracking, and Life in a North Dakota Boomtown* (New York: St. Martin's Press, 2017), 106.

6   Jeff Brady, "After Struggles, North Dakota Grows Into Its
      Ongoing Oil Boom," *All Things Considered*, National Public
      Radio, November 23, 2018, https://www.npr.org/2018/11/23
      /669198912/after-struggles-north-dakota-grows-into-its
      -ongoing-oil-boom.

7   "Petroleum & Other Liquids," U.S. Energy Information
      Administration, accessed March 2020, https://www.eia.gov/dnav
      /pet/hist/rwtcD.htm.

8   Will Englund, "Bravado, dread and denial as oil-price collapse
      hits the American fracking heartland," *Washington Post*,
      March 26, 2020, https://www.washingtonpost.com/business
      /2020/03/25/bravado-dread-denial-oil-price-collapse-hits
      -american-fracking-heartland/?utm_campaign=wp_the_energy
      _202&utm_medium=email&utm_source=newsletter&wpisrc=nl
      _energy202.

9   Russell Gold, *The Boom: How Fracking Ignited the American Energy
      Revolution and Changed the World* (New York: Simon and Schuster,
      2014), 303.

15: THE CORE OF THE CORE

1   Amy Dalrymple, "Discarded RVs a nuisance for Bakken salvage
      yard," *Bismarck Tribune*, August 19, 2016, https://bismarcktribune
      .com/news/state-and-regional/discarded-rvs-a-nuisance-for
      -bakken-salvage-yard/article_aa448dbf-40ef-5bd2-8e18-cdc
      54d51707b.html.

2   Jeff Brady, "After struggles, North Dakota grows into its
      ongoing oil boom," *All Things Considered*, National Public Radio,
      November 23, 2018, https://www.npr.org/2018/11/23/669198912
      /after-struggles-north-dakota-grows-into-its-ongoing-oil-boom.

3   Gregory Zuckerman, *The Frackers* (New York: Penguin Group,
      2013), 147.

4   Zuckerman, *The Frackers*, 154.

5   Dobb, "The New Oil Landscape."

6   Bryan Gruley, "The Man Who Bought North Dakota," *Bloomberg Businessweek*, January 19, 2012, https://www.bloomberg.com /news/articles/2012-01-19/the-man-who-bought-north-dakota.

7   Louisa Kroll and Kerri Dolan, eds.,"The Forbes 400: The Definitive Ranking of the Wealthiest Americans," *Forbes*, October 2, 2019, https://www.forbes.com/forbes-400 /#4d25cbd27e2f.

8   Rivka Galchen, "Weather Underground: The arrival of man-made earthquakes," *The New Yorker*, April 6, 2015, https://www .newyorker.com/magazine/2015/04/13/weather-underground.

9   Joe Wertz, "Oklahoma Earthquake Was Largest Linked to Injection Wells, New Study Suggests," *State Impact Oklahoma*, National Public Radio, March 26, 2013, https://stateimpact.npr .org/oklahoma/2013/03/26/oklahoma-earthquake-was-largest -linked-to-injection-wells-new-study-suggests/.

10   Galchen, "Weather Underground."

11   Mike Soraghan, "Former Okla. seismologist confirms pressure, conflicts of interest in TV interview," *Energywire*, December 9, 2015, https://www.eenews.net/stories/1060029180/.

12   Mike Soraghan, "Hamm says he wasn't pressuring Okla. scientist, but seeking information," *E&E News*, May 11, 2015, https:// www.eenews.net/stories/1060018280.

13   Benjamin Elgin, "Oil CEO Wanted University Quake Scientists Dismissed: Dean's E-Mail," *Bloomberg*, May 15, 2015, https://www.bloomberg.com/news/articles/2015-05-15 /oil-tycoon-harold-hamm-wanted-scientists-dismissed-dean-s -e-mail-says.

14   Bailey Lewis, "Earthquakes continue to decrease in Oklahoma for third straight year," *OUDaily*, March 14, 2019, http://www .oudaily.com/news/earthquakes-continue-to-decrease-in -oklahoma-for-third-straight-year/article_00cefc9c-467f-11e9 -b984-2bebe425ee8e.html.

15   Gruley, "The Man Who Bought North Dakota."

16: A CONVERSATION WITH TEDDY

1   Doris Kearns Goodwin, *The Bully Pulpit: Theodore Roosevelt, William Howard Taft, and the Golden Age of Journalism* (New York: Simon and Schuster, 2013), 286.

2   Quoted in "Refuge of the American Spirit," video produced for Theodore Roosevelt National Park by Theodore Roosevelt Nature and History Association, Medora, North Dakota, 2013.

3   Theodore Roosevelt, *Ranch Life and the Hunting-Trail* (New York: The Century Co., 1888), 37.

4   Patricia Cohen and Maggie Astor, "For Democrats Aiming Taxes at the Superrich, 'the Moment Belongs to the Bold,'" *New York Times,* February 8, 2019, https://www.nytimes.com/2019/02/08 /business/democratic-wealth-tax-warren-sanders-ocasio-cortez .html?searchResultPosition=1.

5   Rebecca Beitsch, "Republicans form conservation caucus to take on environment, climate change," *The Hill,* July 10, 2019, https:// thehill.com/policy/energy-environment/452399-republicans-form -conservation-caucus-to-take-on-environment-climate.

6   Probably the most famous of Roosevelt's critics was Mark Twain, especially regarding Roosevelt's imperialist leanings. "Mr. Roosevelt is the most formidable disaster that has befallen the country since the Civil War—but the vast mass of the nation loves him, is frantically fond of him, even idolizes him. This is the simple truth. It sounds like a libel upon the intelligence of the human race, but it isn't; there isn't any way to libel the intelligence of the human race." Autobiographical dictation September 1, 1907. Published in *Autobiography of Mark Twain, Vol. 3* (Berkeley: University of California Press, 2015), 136.

7   Edmund Morris, *The Rise of Theodore Roosevelt* (New York: Random House, 1979), 477.

8   "Theodore Roosevelt and Conservation," Theodore Roosevelt National Park, National Park Service, updated November 16, 2017, https://www.nps.gov/thro/learn/historyculture/theodore -roosevelt-and-conservation.htm.

9 Morris, *The Rise of Theodore Roosevelt*, 325.

10 Theodore Roosevelt, "Conservation as a National Duty" speech, White House Conference of Governors, Washington DC, May 13, 1908, http://voicesofdemocracy.umd.edu/theodore -roosevelt-conservation-as-a-national-duty-speech-text/.

11 H. W. Brands, *American Colossus: The Triumph of Capitalism, 1865–1900* (New York: Anchor Books, 2010), 611.

12 *Banking on Climate Change: Fossil Fuel Finance Report Card 2019* (San Francisco: Rainforest Action Network, April 2019), https:// www.ran.org/wp-content/uploads/2019/04/BOCC_2019 _SUMMARY_vUS-1F.pdf.

13 David McCulloch, *Mornings on Horseback: The story of an extraordinary family, a vanished way of life and the unique child who became Theodore Roosevelt* (New York: Simon and Schuster, 2001), 350.

14 McCulloch, *Mornings on Horseback*, 349.

15 James MacPherson, "Planned refinery by national park hurt by funding, lawsuits," *Washington Times*, January 10, 2020, https:// www.washingtontimes.com/news/2020/jan/10/planned -refinery-by-national-park-hurt-by-funding-/.

16 April Baumgarten, "180 wells OK'd near Little Missouri River— North Dakota 'treasure' at risk, conservationists say," *West Fargo Pioneer,* May 24, 2019, https://www.westfargopioneer.com /news/1350658-180-wells-OKd-near-Little-Missouri-River —North-Dakota-treasure-at-risk-conservationists-say.

17 Terry Tempest Williams, *The Hour of Land: A Personal Topography of America's National Parks* (New York: Sarah Crichton Books, 2016), 64.

18 Williams, *The Hour of Land*, 56.

17: A CERTAIN RELENTLESSNESS

1 "Energy and the economy," Natural Resources Canada, Government of Canada, 2018 figures retrieved October 1, 2019, https://www.nrcan.gc.ca/energy-and-economy/20062#L4.

2   Charles Bowden, "The Emptied Prairie," *National Geographic,*
     January 2008, 1–20.

3   Andy Skuce, "Alberta's bitumen sands: 'Negligible' climate effects,
     or the 'Biggest carbon bomb on the planet'?" *Skeptical Science* blog,
     April 28, 2012, https://skepticalscience.com/SW12.html.

4   K. Hayhoe, D. J. Wuebbles, D. R. Easterling, D. W. Fahey, S.
     Doherty, J. Kossin, W. Sweet, R. Vose, and M. Wehner, 2018.
     "Our Changing Climate," in *Impacts, Risks, and Adaptation in
     the United States: Fourth National Climate Assessment, Volume II*
     (Washington DC: US Global Change Research Program, 2018),
     79. In addition, Katharyn Hayhoe of Texas Tech University has a
     superb lay-overview of this topic at https://www.youtube.com
     /watch?v=k5_zpjerQFo&list=PLwNT4Fr0_4CRlYFj3hZPVSaV
     YZfk9YQM4&index=26&t=0s.

5   *IPCC Special Report on the Ocean and Cryosphere in a Changing
     Climate: Summary for Policymakers* (Geneva: Intergovernmental
     Panel on Climate Change, 2019), https://www.ipcc.ch/site/assets
     /uploads/sites/3/2019/09/SROCC_SPM_HeadlineStatements.pdf.

6   Hayhoe et al., "Our Changing Climate," 74.

7   Rich, "Losing Earth."

8   Tik Root, Lisa Friedman, and Hiroko Tabuchi, "Following the
     Money That Undermines Climate Science," *New York Times*, July
     10, 2019, https://www.nytimes.com/2019/07/10/climate/nyt
     -climate-newsletter-cei.html.

9   Turner, *The Patch*, Kindle edition.

10   Harold Hamm asked about climate, talks about California
     irrigation at 1:01:19 of https://www.youtube.com/watch?time
     _continue=3679&v=rbmmR0bXZNQ; Hamm on climate and
     Islamic terrorism at 3:34 of https://www.youtube.com
     /watch?time_continue=215&v=C5f5-ciH1bk. Clips courtesy of
     https://www.desmogblog.com/harold-hamm.

11   Justin Wm. Moyer, "The Legacy of Oil Executive Christophe de
     Margerie, Dead in Freak Moscow Plane Crash," *Washington Post,*

October 21, 2014, https://www.washingtonpost.com/news
/morning-mix/wp/2014/10/21/the-legacy-of-oil-executive
-christophe-de-margerie-killed-in-freak-moscow-crash-of-his
-corporate-jet/.

12 Brett Samuels, "Trump absent from G-7 session on climate," *The
    Hill*, August 26, 2019, https://thehill.com/homenews
    /administration/458785-trump-absent-from-g-7-session-on
    -climate.

13 Rich, "Losing Earth."

14 Andrew Grant and Mike Coffin, *Breaking the Habit: Why none of
    the large oil companies are 'Paris-aligned', and what they need to do to
    get there* (London: Carbon Tracker Initiative, 2019), https://www
    .carbontracker.org/reports/breaking-the-habit/.

15 Matt O'Brien, "The two big reasons there really might be a
    recession in 2020," *Washington Post*, November 20, 2018, https://
    www.washingtonpost.com/business/2018/11/20/two-big-reasons
    -there-really-might-be-recession/?utm_term=.8c0eaf021d90.

16 Larry Fink, "A Fundamental Reshaping of Finance," BlackRock,
    January 13, 2020 https://www.blackrock.com/corporate
    /investor-relations/larry-fink-ceo-letter.

17 Eugene Linden, "How Scientists Got Climate Change So
    Wrong," *New York Times*, November 8, 2019, https://nyti
    .ms/34D2av8.

18 Tony Seba, "Tony Seba: Clean Disruption - Energy &
    Transportation," presentation, Colorado Renewable Energy
    Society, Boulder, CO, June 9, 2017, https://www.youtube.com
    /watch?v=2b3ttqYDwF0.

19 Allen et al., *Global Warming of 1.5°C*, SPM-15.